# RIVER ROAD VISTAS

# RIVER ROAD VISTAS

## A Journey Along the Rio Grande

By

William MacLeod

First Printing, May 2008

Printed and Bound in the United States

Cover Photo: “Santana Basin from Big Hill” by William MacLeod

Publishers Cataloging in Publication Data
MacLeod, William
River Road Vistas: a journey along
the Rio Grande/ by William MacLeod
p. cm.
Includes glossary and index
ISBN 978-0-9727785-3-5
1. Geology – Presidio County Region (Tex.). I. Title.
QE168.B5 M24 2005
557.64-dc22

TEXAS GEOLOGICAL PRESS
P.O. Box 967
Alpine, Texas 79831

*For Martha*

# 1 Introduction

Previous page Three Dike Hill, photographed on the River Road 42.5 miles west of Study Butte.

For a description of strata on the mountain, see page 112.

River Road Vistas takes you along a 230-mile loop from Alpine down to the Rio Grande and back. The loop runs south from Alpine for 77 miles to Study Butte, at the western entrance to Big Bend National Park. Turning right along the Rio Grande valley, the road continues for 66 miles passing the ghost town of Terlingua, site of an old mercury mine, and through the recently expanded Lajitas Resort.

The heart of the journey is the 30-mile stretch between Lajitas and Redford, the River Road, where the river follows a narrow corridor between mountainous terrain formed by the violent eruptions of volcanoes to the north and south that were active millions of years ago. Mesas and ridges rise as much as 1,500 feet above the highway and the river in places cuts through canyons that are up to 1,000 feet deep. This is the most spectacular scenery in Texas.

At Presidio, the road leaves the Rio Grande and climbs north through barren terrain to the Chinati Mountains and the other ghost town on the route, the old silver-mining settlement of Shafter. Beyond Shafter, it ascends gradually to a summit nearly 3,000 feet above Presidio before descending to the vast plain around Marfa, 60 miles north of Presidio. Marfa, an old military and ranching market town, now has a substantial colony of artists who have been attracted by the open spaces and the wonderful Texas light. The final 25 miles travels east to Alpine through the Paisano Volcano caldera filled with collapsed and chaotically jumbled rocks. This is the best close-up view of a volcano in Texas.

The full loop takes the better part of a day to complete, including stops for lunch and photography. The roads are two-lane blacktops in good condition, although hilly and winding in places. Food and refreshments are available in Alpine, Study Butte, Terlingua, Lajitas, Presidio and Marfa. Gasoline is available in Alpine, Study Butte, Presidio and Marfa. Conditions are mild and pleasant along the route for most of the year. In summer, however, high afternoon temperatures are the norm along the Rio Grande, which is only 2,500 feet above sea level, compared to 4,500 feet at Alpine and 4,800 feet at Marfa. At that time of year you should plan on traveling from Study Butte to Presidio in the morning.

Incidentally, the River Road between Lajitas and Presidio should not to be confused with the dirt road between Castolon and Tornillo Bridge in Big Bend National Park, also called the River Road.

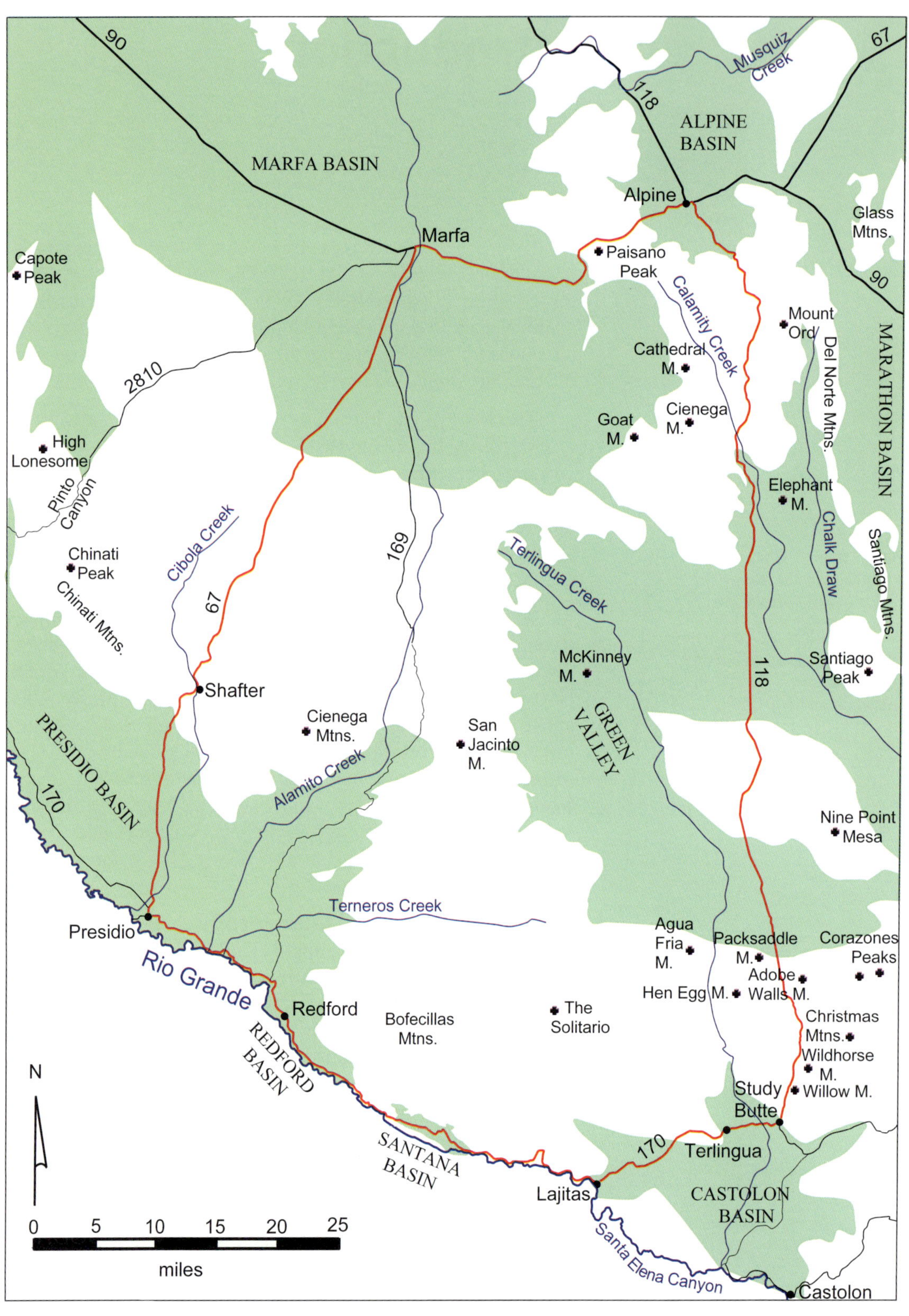

90
Musquiz Creek
67
118
ALPINE BASIN
MARFA BASIN
Alpine
Glass Mtns.
Marfa
Paisano Peak
Capote Peak
90
Calamity Creek
Mount Ord
Del Norte Mtns.
MARATHON BASIN
2810
Cathedral M.
Goat M.
Cienega M.
High Lonesome
Pinto Canyon
Elephant M.
Chalk Draw
Santiago Mtns.
Cibola Creek
169
Terlingua Creek
Chinati Peak
Chinati Mtns.
67
McKinney M.
118
Santiago Peak
Shafter
Cienega Mtns.
San Jacinto M.
GREEN VALLEY
PRESIDIO BASIN
170
Alamito Creek
Nine Point Mesa
Terneros Creek
Presidio
Agua Fria M.
Packsaddle M.
Corazones Peaks
Rio Grande
Adobe Walls M.
Hen Egg M.
Redford
Bofecillas Mtns.
The Solitario
Christmas Mtns.
REDFORD BASIN
Wildhorse M.
N
Study Butte
Willow M.
SANTANA BASIN
170
Terlingua
Lajitas
CASTOLON BASIN
0
5
10
15
20
25
miles
Santa Elena Canyon
Castolon

## Topography

The River Road area is mountainous, certainly by Texas standards. Some mountains are the result of compression of the Earth's crust. These include the Santiago Mountains to the east of Green Valley which rise up to 6,000 feet above sea level. The Glass Mountains to the east of Alpine are likewise due to compression which created a dome around Marathon. The dome is now mostly eroded away except for the Glass and Del Norte Mountains on its rim.

Other mountains are the eroded remains of volcanoes or volcanic flows. The Chinati Mountains northwest of Presidio are in this category as are the Christmas Mountains near Study Butte and the Bofecillos Mountains between Lajitas and Presidio. Intrusions make up a third category of mountains. Formed when bodies of magma solidified underground, the intrusions were later uncovered by erosion. Examples include Twin Peaks at Alpine, Elephant Mountain and Santiago Peak in Green Valley, and Paisano Peak near Marfa.

Between the mountainous areas are basins, shown in green on the map opposite. Basins are generally flat-bottomed and are produced by erosion, faulting, or a combination of both; some are linked together by canyons or draws.

### River Road Topography

**Left** The River Road loop is shown in red. Flat-bottomed topographic basins are shown in green. Mountainous or hilly areas are in white.

The faulted Castolon Basin south of Terlingua includes erosion up Terlingua Creek. Several small basins are found along the Rio Grande such as the Santana Basin south of the Bofecillos Mountains and another around Redford. Both are underlain by grabens. The Presidio Basin overlies a graben south and west of the Chinati Mountains, and has been extended by erosion up Alamito Creek.

The Marfa Basin, the largest in the area, stretches as a series of flats from Goat Mountain southwest of Alpine to the foothills of the Guadalupe Mountains on the border of New Mexico, 120 miles to the northwest. Apart from a few low lava-capped hills and shallow creek valleys, the basin is relatively flat and lies from 4,500 to 5,000 feet above sea level. The basin is underlain by a series of *grabens* in which strata have been down-dropped by faults on one or both long sides (see diagram on page 38). The grabens filled with debris from the higher ground around them as they sank, usually a mixture of silt, sand and gravel called *graben fill* by geologists.

The small Alpine Basin is at least partly the result of faulting along its southwestern and northeastern boundaries. Between 4,300 and 4,500 feet in altitude, the basin is covered by a thin (0-50 feet) layer of alluvium and connects to the Marathon Basin through a narrow fault-bounded pass between the Glass and Del Norte Mountains. South of Alpine, the Green Valley basin developed mainly through erosion by the Terlingua, Calamity and Chalk Draw Creeks. The basin slopes gently towards the Rio Grande from about 4,000 feet above sea level in the north to 3,500 feet in the south.

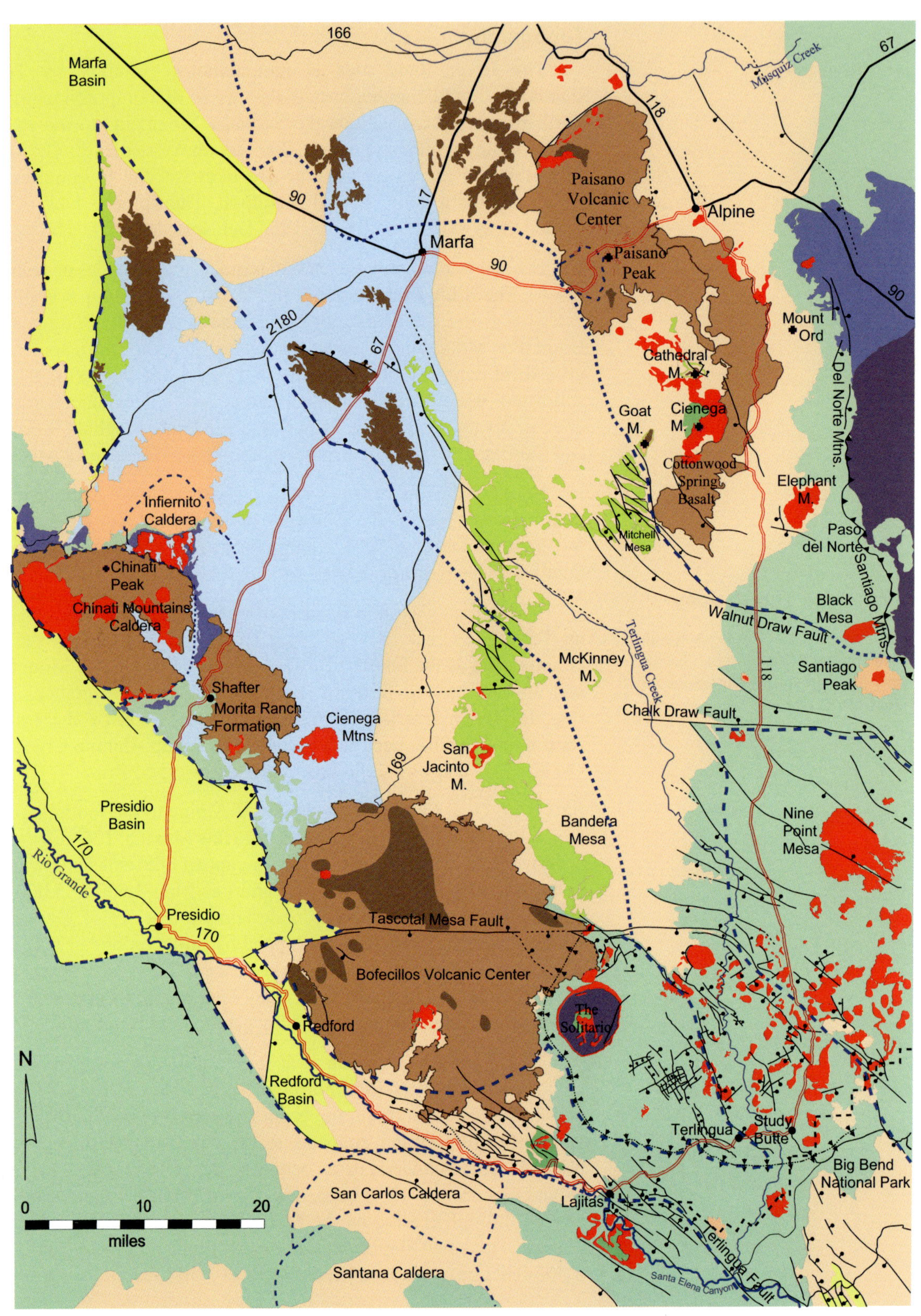
Marfa Basin
166
Musquiz Creek
67
118
Paisano Volcanic Center
90
17
Marfa
Alpine
Paisano Peak
90
90
2180
67
Mount Ord
Cathedral M.
Del Norte Mtns.
Goat M.
Cienega M.
Cottonwood Spring Basalt
Elephant M.
Infiernito Caldera
Mitchell Mesa
Paso del Norte
Santiago Mtns.
Chinati Peak
Chinati Mountains Caldera
Black Mesa
Walnut Draw Fault
Terlingua Creek
McKinney M.
118
Santiago Peak
Shafter
Morita Ranch Formation
Chalk Draw Fault
Cienega Mtns.
San Jacinto M.
169
Presidio Basin
Bandera Mesa
Nine Point Mesa
170
Rio Grande
Presidio
170
Tascotal Mesa Fault
Bofecillos Volcanic Center
The Solitario
Redford
N
Redford Basin
Terlingua
Study Butte
Big Bend National Park
San Carlos Caldera
Lajitas
0
10
20
miles
Terlingua Fault
Santana Caldera
Santa Elena Canyon

EXPLANATION

Graben fill

Perdiz Conglomerate

Petan Basalt

Mitchell Mesa Rhyolite

Igneous intrusions

Named volcanic rocks

Other volcanic rocks

Cretaceous strata

Permian strata

Early Paleozoic strata

Normal fault; tick on down-thrown side; dashed where inferred

Thrust fault; barbs on overriding plate

Caldera boundary

Rift segment boundaries dotted where inferred

Terlingua Monocline

## Geology of the Region

The River Road Vistas area is at the eastern edge of the North American Cordillera, a mountainous region created by disruptions in the Earth's crust, disruptions studied in the branch of geology known as *tectonics.*

In the Big Bend, there were five such disruptions in the last 70 million years. The first disruption lasted from 70 to 60 million years ago when the crust was compressed across North America, creating ranges such as the Santiago Mountains and Sierra del Carmen east of the Alpine-Study Butte road.

In the second disruption, from 47 to 27 million years ago, massive volcanic flows erupted from centers such as those of the Paisano area near Alpine, the Chisos Mountains in the Big Bend, the Bofecillos Mountains along the Rio Grande, and the Chinati Mountains south of Marfa.

Another period of compression across the area followed about 32 million years ago or later, creating minor folding which you can see south of Alpine, in Green Valley and around Lajitas.

The fourth disruption began about 25 million years ago when the continent began to be uplifted along the Rocky Mountains. Thus Cretaceous limestones in the Alpine area are now some 4,500 feet above sea level, whereas, when they first developed, they were below sea level. The uplift may still be continuing.

About the same time, 27 million years ago, a rift zone began to develop along the crest of the continent, a zone known as the Rio Grande Rift and now followed by the Rio Grande from Colorado to the Big Bend. In this zone, the Earth's crust was stretched or extended and broke into fault blocks. Some blocks stayed high while others sank down and were covered with erosional debris. Basalt lavas erupted through the faults and spread over the landscape in thin flows from about 25 to 18 million years ago. South of Socorro, New Mexico, the zone widened into a series of parallel segments which continued as far south as the Big Bend in the Rio Grande.

The River Road loop crosses one of the segments in the rift zone, the Salt Basin rift segment, south of Alpine and south of Marfa and follows the main Rio Grande rift segment from Study Butte to Presidio. Rifts and their segments are discussed in detail in Chapter 2 (page 39).

With all these events in its history, it is not surprising that the River Road Loop has an astonishing variety of spectacular scenery.

**Satellite Photograph of the River Road Loop**

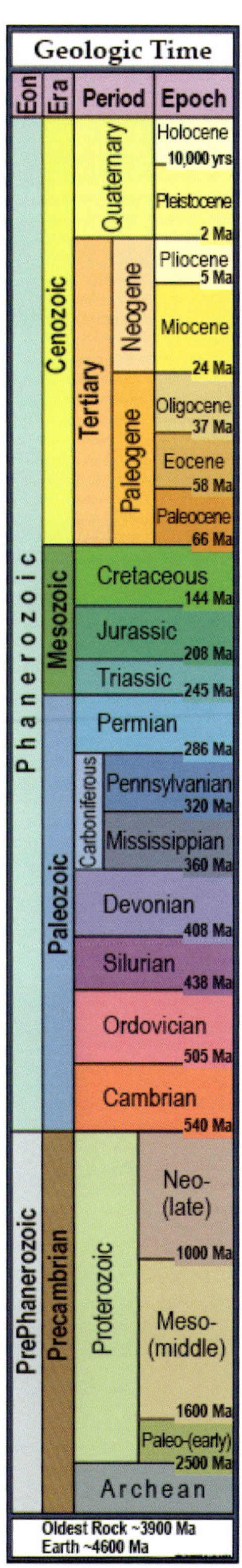


Books on geology always have to begin with a table such as the one to the left to introduce the general reader to the geological time scale and the terminology involved. The time scale, also called the *geological column,* begins with the creation of the Earth, about 4.6 billion years ago and ends at the present day.

The basic unit of time is one million years or *m.y.* Events are described as having occurred or taken place so many millions of years ago, abbreviated as *Ma.*

The time scale is divided into *eras*, Precambrian, Paleozoic, Mesozoic and Cenozoic, which are in turn divided into *periods.* Many of the names come from the place names where the rocks of the period were first described such as Pennsylvanian from Pennsylvania. Periods are further divided into epochs. The rocks that were created in a period or epoch are tabulated into *formations*, bodies of rock that can be identified in the field by their physical characteristics and position in the geological time scale. A formation is sometimes subdivided into *members* or combined with other formations into a *group.*

Turning to the geology of the River Road Vistas area, the rift segments crossing the area are outlined in dashed lines where boundaries are known, dotted lines where boundaries are uncertain or indistinct. The lines are blue on the geological map on page 12 and yellow on the satellite photograph opposite. The Salt Basin segment comes down from the northwest to the Chalk Draw Fault where it is bifurcated by a block that includes the Christmas and Chisos Mountains. The Rio Grande segment runs along the Rio Grande through Presidio and Lajitas to join the Salt Basin segment at Study Butte.

The region is underlain about equally by sedimentary rocks of the Cretaceous period, volcanic rocks of the Tertiary period and alluvial sediments also of the Tertiary period.

The Cretaceous sedimentary rocks are mostly limestones. They began as lime mud, a mud composed of minute crystals of aragonite (a mineral, chemical composition $CaCO_3$) derived from the breakdown of calcareous algae skeletons and possibly chemical precipitation. The crystals sank to the ocean floor and in time hardened into limestone. These rocks are the same age as the limestones in the Texas Hill Country.

Paleozoic sedimentary rocks are exposed in a few places along the route, in the Solitario, for example, and around Shafter. None can be seen from the highway, however.

Volcanic rocks of the main volcanic period, 47 to 27 Ma, are shown as Volcanic Rocks and Named Volcanic Rocks on the geological map. Petan Basalt erupted from fissures in or near rift zones from 25 to 18 Ma.

**Right** This table gives the common definitions of grain or clast sizes in sediments and the names of the equivalent sedimentary rocks when the sediments become *indurated,* hardened or consolidated by pressure, heat or cementation.

| Clast | Size | Rock when indurated |
|---|---|---|
| clay | less than .00016 inch | mudstone; if with partings, shale |
| silt | between .00016 and .0025 inch in diameter | siltstone |
| sand | between .0025 and .08 inch in diameter | sandstone |
| gravel | greater than .08 inch in diameter | conglomerate |

Two alluvial sedimentary formations, accumulations of silt, sand and gravel deposited by streams or rivers, are shown on the map. The older formation is the Perdiz Conglomerate found west and south of Marfa, a thick accumulation of boulders, gravel, sand and silt, which is very lightly cemented. It developed from erosion of the Chinati Mountains and other high ground. The other alluvial category shown on the geological map is graben fill, introduced on page 11.

### The Satellite View

The satellite photograph on page 14 shows the entire River Road area as seen from space with the roads along the loop shown in red. Other roads are in black.

Segments of the Rio Grande Rift are outlined with yellow lines, dashed where they can be placed with some certainty, dotted where uncertain. The lightly-vegetated silvery gray Presidio Basin north of Presidio in the Rio Grande segment is particularly striking as is the more heavily-vegetated yellowish green Marfa Basin northwest of Marfa in the Salt Basin segment.

Basins not in rift segments also show up well, the yellow green Alpine Basin north of Alpine and the light gray eastern section of Green Valley south of Elephant Mountain underlain by limestones. Similar light gray limestones underlie the Castolon Basin southwest of Study Butte. The darker western part of Green Valley is mostly underlain by volcanic tuffs.

Several rock types can be identified by their colors on the photograph. Basalts, which are dark-gray to black on the ground, show up as dark areas like the Bofecillos Mountains and the areas south of Alpine, Marfa and Kokernot Mesa. The light-colored Mitchell Mesa Rhyolite welded tuff is the dark yellow band running from Bandera Mesa to Mitchell Mesa. Large intrusions such as Nine Point Mesa and Elephant Mountain are brown on the photograph. The many small intrusions north of Study Butte are dark brown or black.

You can compare the photograph with the geological map of the area to see other parallels.

# 2 Alpine to Study Butte

The first section of the River Road loop is the Alpine to Study Butte road, 76.5 miles along an excellent two-lane road. The trip takes about two hours, including time for photographs and other stops. Mileages in this chapter are measured from the flashing red light at the intersection of US 90 and TX 118 South. Strata along the route are described in the table on page 28.

The highway begins in alluvium of the Alpine Basin and 10 miles south of town climbs Big Hill onto a volcanic plateau. The plateau is underlain by Cottonwood Spring basaltic lavas over tuffs and lavas of the Pruett Formation.

At Mile 33.5 the road descends the Walnut Draw Fault escarpment on to flaggy limestones of the Cretaceous Boquillas Formation which underlie the road across Green Valley for the next 30 miles.

In its final 15 miles, the Alpine-Study Butte highway cuts through Upper Cretaceous claystones of the Pen and Aguja Formations between enormous numbers of intrusions, large and small as it descends to Study Butte, 1,500 feet below Green Valley.

## Volcanoes and Volcanic Rocks

Visitors to the area find volcanoes fascinating and want to know where they can be seen. The answer, of course, is that they are all around us and the violent consequences of their eruptions are everywhere. I begin this chapter by explaining why they are here in the first place and then identify the places where you can see them.

Volcanoes in the Trans-Pecos of Texas developed in two episodes, the greatest between 47 and 27 Ma, with a secondary episode between 25 and 18 Ma. The Trans-Pecos volcanic field is part of an arc of volcanic rocks roughly parallel to the west coast of northern Mexico and the southern United States (map on page 18).

The volcanic arc developed because the North American plate, the section of the Earth's crust on which North America sits, began over-riding the Farallon plate at the west coast of the continent. As the Farallon plate descended into the Earth's interior below North America, it carried with it vast amounts of water and water-saturated sediments from the sea floor. Once this water-laden material reached

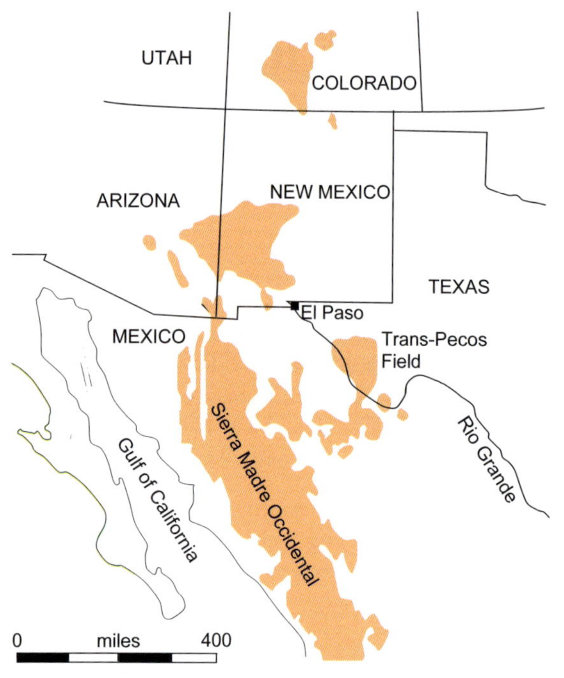


## Tertiary Volcanism

**Left** During the middle Tertiary period, subduction of the oceanic Farallon Plate beneath the west coast of North America created a volcanic arc along western North America. The remaining outcrops of the arc are shown in this map.

The River Road is in the Trans-Pecos volcanic field.

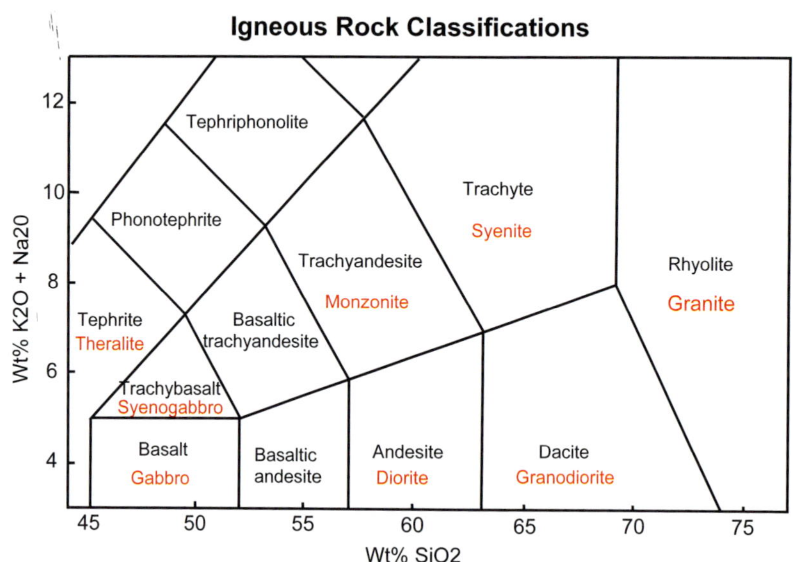


**Above** One of the ways to classify igneous rocks is by comparing their silica content ($SiO_2$) to the sum of their sodium and potassium oxides ($Na_2O$+$K_2O$). In this diagram, fine-grained rocks, those with crystal diameters .25mm or less, are named in black and their coarser-grained equivalents in red.

Most of the rocks created during the main volcanic phase in West Texas (between say 38 and 28 Ma) are high in silica (70-75%), in the rhyolite and trachyte sectors.

Rocks produced during the Petan Basalt period (23-18 Ma) are mainly basalts and trachybasalts.

75 miles or so below the Earth's surface, it began to melt and the molten rock or *magma* rose in the crust under the force of gravity.

Once in the crust the magma formed pools or reservoirs of molten rock in *magma chambers*, typically 20 to 60 miles underground. Eventually, magma reached the Earth's surface through vents or fissures and created volcanic eruptions. In some cases, erosion has destroyed evidence of the nature of the eruptions but many volcanic vents or chimneys have been identified in the area, none of which can be seen from the highway, unfortunately. Other volcanic flows came from multiple fissures or cracks in the Earth's surface. One particularly good example is the Paisano Volcano near Alpine, where erosion has uncovered dikes, vertical rock bodies that began as magma in fissures. Some 1,000 of these dikes have been identified in the 450-square mile area of the volcano (see photograph on page 173).

Some magma, especially the magma first erupting, contained water vapor, and exploded into clouds of small droplets on reaching the surface. The droplets very quickly solidified into volcanic glass and depending on the strength of the explosions, fell to the ground locally or rose up into the atmosphere and were carried off by winds, eventually falling to Earth as *volcanic ash*.

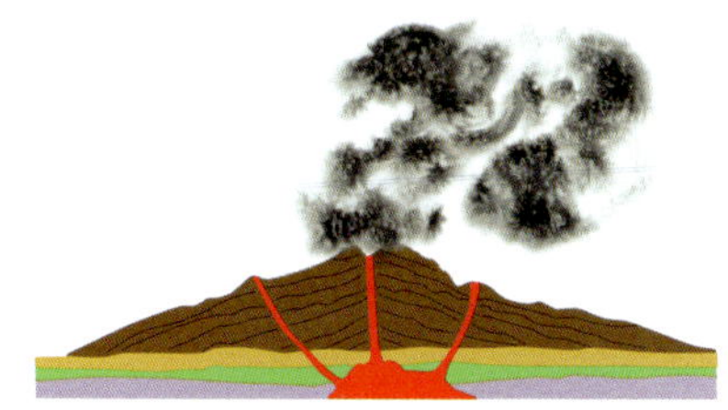

**Stratovolcano**

A stratovolcano such as Mount Ranier gradually builds up a cone through erupting multiple lava flows from one or more vents. This illustration shows three vents.

Eventually, the volcanic ash deposits became cemented into a soft rock called *tuff* by minerals such as calcite, silica and iron oxide carried in by groundwater. In some cases, the hot ash partially re-melted as it accumulated to create a rock called *welded tuff*. All degrees of welding, from rocks that are only slightly welded to rocks almost identical to lava, are found in the River Road area.

Magma that lacked water vapor or other volatile matter oozed out as lava on to the Earth's surface. In the Trans-Pecos, most igneous rocks, a collective term for volcanic and intrusive rocks, were rhyolites or trachytes with some basalts and trachybasalts (the diagram opposite classifies igneous rocks). The rhyolites and trachytes were viscose, even when hot, having the consistency of oatmeal, and tended to form thick, stumpy flows called *lava domes* near their vents or fissures (a lava dome can be seen in the photograph on page 102). Other volcanoes were probably conical *stratovolcanoes*, similar to those in the Pacific Northwest, Mount Ranier or Mount Saint Helens for example, which are being created today in an active volcanic arc.

**Mount Ranier, Seattle WA**

In places, a circular or oval depression on the surface known as a *caldera* developed as the roof of a magma chamber collapsed when all magma had erupted. Some can be seen in satellite photographs; others can only be recognized from variations in thickness of rock layers on the surface. The River Road loop passes alongside the Chinati Mountains Caldera and cuts through the Paisano Caldera.

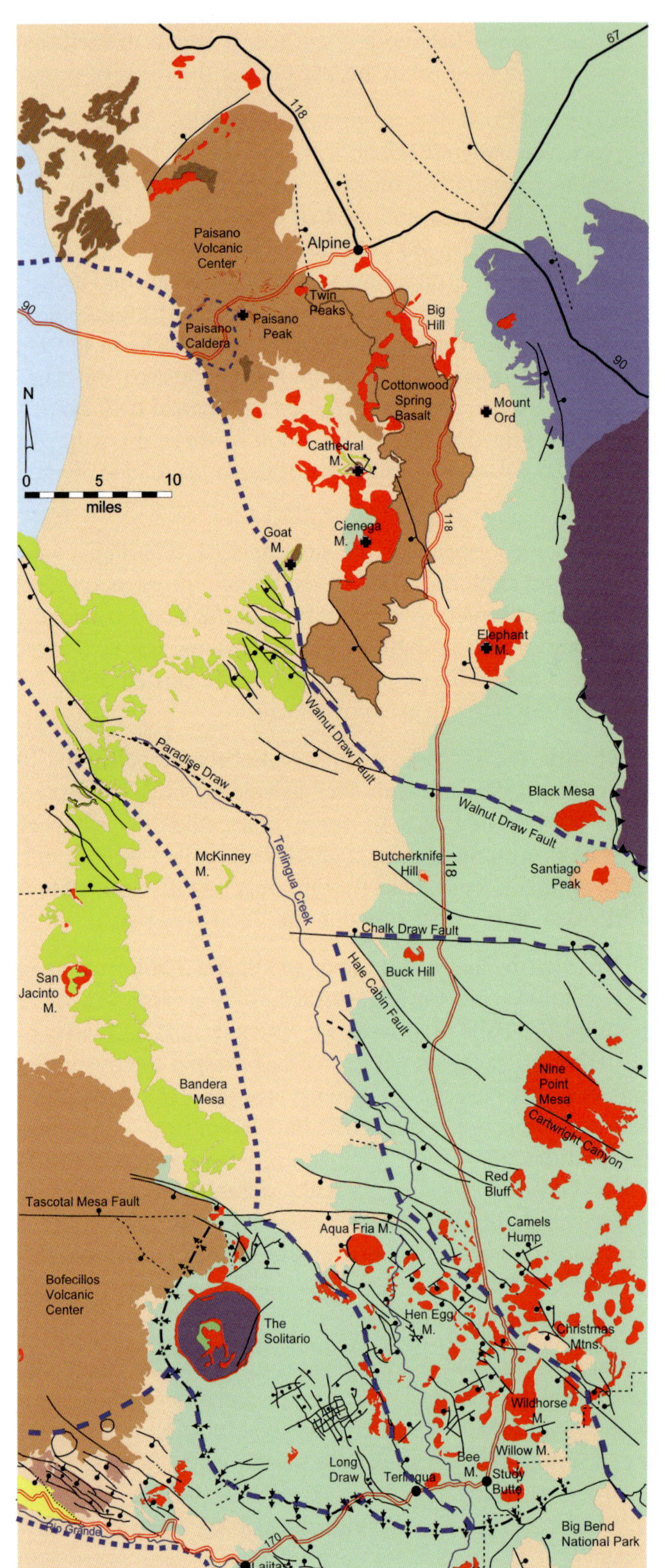


## EXPLANATION

- Perdiz Conglomerate
- Petan Basalt
- Mitchell Mesa Rhyolite
- Igneous Intrusions
- Named Volcanic Rocks
- Other Volcanic Rocks
- Cretaceous Strata
- Permian Strata
- Early Paleozoic Strata
- Normal fault; tick on down-thrown side; dashed where inferred
- Thrust fault; barbs on overriding plate
- Rift segment boundaries dotted where inferred
- Terlingua Monocline

**Geology: Alpine to Study Butte**

The oldest volcanic rock in the River Road area is a thick blanket of tuff that occurs from the Santiago Mountains to Marfa and down across the Rio Grande into Mexico. The rock is called the Chisos Formation down by the Rio Grande and the Pruett Formation around Alpine.

Lava eruptions in Texas began at the Christmas Mountains near Study Butte about 47 Ma and picked up volume about 36.8 Ma, when the Crossen lava of the Pruett Formation, and its equivalent in the Davis Mountains, the Star Mountain Formation, began spreading across the landscape from the Davis Mountains down to Elephant Mountain. Volcanic eruptions continued west of Alpine until 36.3 Ma in the Paisano Volcano. At roughly the same time, eruptions along the Rio Grande produced tuffs and lavas of the Chisos Formation which we will see along the river from Study Butte westwards. Later eruptions created the Chinati volcanic field ending at 32 Ma and Bofecillos volcanic field ending at 27 Ma.

A second episode of volcanic activity occurred from 25 to 18 Ma when basalt lavas erupted up through faulting in rift zones. These basalts are called Petan Basalt on the geological maps.

### Igneous Intrusions

Magma that solidified underground before reaching the surface formed *igneous intrusions,* some of which were later uncovered by erosion. Geologists classify intrusions by shape, and, to some extent, size. A tabular or sheet-like body is called a *sill* if it has been injected parallel to the bedding in its host rock. The shape of an uncovered intrusion depends partly on the shape of the original intrusion, so sills, being parallel to the beds into which they were injected, are horizontal or near horizontal, and often create mesas when uncovered. Elephant Mountain and Nine Point Mesa on the Study Butte-Alpine road are examples.

A *laccolith* is a sill that domes the beds above it, forming a mushroom-shaped body. Laccoliths are common in the area and tend to produce mushroom-shaped mountains. An example is Cienega Mountain south of Alpine. If a tabular-shaped intrusion crosscuts the bedding of its host rock, it is called a *dike*. Dikes are usually nearly vertical. They tend to be harder than lavas and stand up above the surface, an example of differential erosion, often creating ridges with cockscomb or iguana lizard profiles. Many examples can be seen on the Alpine-Marfa road, Lizard Mountain for example or in front of Packsaddle Mountain (see photograph on page 53).

A *plug* is a vertical, pipe-like body, sometimes a volcanic *neck*, or a solidified intrusion feeder. Plugs or volcanic necks are cylindrical bodies that narrow toward their peaks when eroded. A good example is Hen Egg Mountain north of Study Butte (photograph on page 54).

**Satellite Photograph - Alpine Basin**

## The Alpine Basin

**Above** Looking across the western Alpine Basin from Big Hill. The two intrusions on the left skyline are the Haystacks (see page 167). The sharp peak at right on the skyline is Mitre Peak, also an intrusion. Mount Livermore, the highest mountain in the Davis Mountains, is 40 miles away on the skyline to the immediate right of the Haystacks.

**Left** The satellite photograph opposite shows the light yellow-green Alpine Basin bracketed by faults. The eastern edge of the brown volcanic rocks is clearly defined from Bullfrog Mountain to Elam Mountain and beyond.

The large dike, 400 feet high and over a mile long, marked on the map as "Crossen Dike" has been identified as a feeder dike for the enormous Crossen lava, which is found from Last Chance Mesa to Kokernot Mesa, 34 miles to the south.

### The Alpine Basin

Although the Alpine Basin has never been mapped in detail, it appears to be underlain by a shallow graben, at least under its southern part. As marked on the satellite photograph on page 22, a fault zone runs along the cliffs on the west across the opening into Sunny Glen; strata in the basin are 750 feet lower than in the cliffs. On its northeast side, two faults drop strata down from Horse Mountain into the basin. The first of these has a down-throw of 600 feet where it crosses the Orient railway line.

Hancock Hill (4,925 feet), with Sul Ross State University on its flank, rises some 500 feet above the basin and is capped by Crossen lava. Hancock Hill appears to be a block left in its pre-basin position. A fault on the west side of the hill drops strata 850 feet down into the basin. There probably is a similar fault on the east of the hill. Some geologists have speculated that the block remained in place because of an underlying intrusion although no direct evidence of an intrusion has yet been discovered.

### The Alpine Basin

**Above** Looking across the Alpine Basin to the west in the morning light, Twin Peaks dominates the horizon on the left. A level lava bed to its right caps a mesa. Highway 90 runs through the gap between the mesa and the pyramid-shaped hill to its right with the railway line running through the opening on the other side of the hill. Paisano Peak is on the horizon behind the gap.

The pyramid-shaped hill on the far right is on the flank of Sunny Glen, the deepest canyon in the Paisano cliffs.

The Alpine Basin is bordered on the west by 1,300-foot cliffs made up of a succession of Decie Formation rhyolitic and trachytic lavas with underlying tuffs erupted from the Paisano volcanic center west of Alpine (see page 166). Dark outcrops of Cottonwood Spring Basalt appear intermittently along the base of the escarpment. Canyons have developed at intervals in the cliffs, the deepest being Sunny Glen, four miles northwest of town.

The cliffs continue across Cottonwood Spring Basalt outcrops and intrusions to Highway 118 at Big Hill. From Big Hill the basin boundary runs north to Last Chance Mesa along a Crossen lava ridge, a ridge that continues south to Elephant Mountain. The high point on this part of the ridge is Bullfrog Mountain (5,400 feet). North of Highway 90, the ridge diminishes in height with isolated crests such as Cone, Horse and Elam Mountains (4,940 feet) on its eastern edge.

### The Volcanic Plateau

Leaving Alpine on TX 118 South, low rolling hills and swales bottom the basin for the first five miles to Big Hill. Alpine or "A" Hill on the right leaving town is a trachyte intrusion. A number of other

## Strata from Alpine to Study Butte

| Period/Epoch | Formation | Age m.y. | Description |
|---|---|---|---|
| Oligocene-Miocene | Petan Basalt | 25-18 | Mafic flows, dark greenish gray to brownish gray; up to 510 ft. thick near Marfa. |
| | Tascotal | ~29 | Sandy rhyolitic tuff, coarse-grained, sandstone, and pebble conglomerate of mafic lavas and fine-grained syenite; light gray to yellow; up to 462 ft. thick on Cathedral Mountain, thins northward. |
| | Mitchell Mesa Rhyolite | 31.5 | Cliff-forming ash flow, moderately to densely welded; weathers dark reddish-gray to black; up to 150 ft. thick. |
| | Duff | 37-35 | Rhyolite tuff with lenticular beds of sandstone, breccia and conglomerate, and thin basaltic lava flows; up to 1,500 ft. thick; combined with tuff of the lower Pruett Formation to make up the Devil's Graveyard Formation south of McKinney Mountain. |
| Oligocene | Cottonwood Spring Basalt | ~36 | Mafic flows; upper part of flows vesicular, amygdaloidal, reddish gray to reddish brown; middle part massive, grayish black to dark greenish black; lower part commonly flow breccia; up to 300 ft. thick. |
| Eocene-Oligocene | Pruett | 46.5 at base in south; 38 at base in north | Mostly tuff, some tuffaceous sandstone, conglomerate breccia and tuffaceous non-marine limestone; tuff grayish white, bluish white, greenish gray, brownish gray, brown, pink and red; erodes to low rounded hills; up to 800 ft. thick. The following members occur in the upper part of the formation:<br>*Potato Hill Andesite:* Plagioclase phenocrysts in fine-grained reddish-brown or coarse-grained grayish brown groundmass; upper half flow breccia, lower half massive, vesicular; about 20-40 ft. thick.<br>*Sheep Canyon Basalt:* At least four flows, fine- to medium-grained, dark to greenish black, up to 235 ft. thick.<br>*Crossen Member:* Massive porphyritic quartz trachyte and rhyolite, stubby feldspar phenocrysts in fine- to medium-grained grayish- to reddish-brown groundmass; hard, brittle, weathers to rusty brown, pitted surfaces; up to 265 ft. thick. |
| Upper Cretaceous | Aguja | 98-70 | Claystone, green and gray with thin lignite and coal beds interbedded with light brown to white sandstones; grading from marine claystone in lower part to continental claystones and sandstones in upper part; vertebrate fossils, petrified wood common; 600 ft. thick at maximum. |
| | Pen | | Mostly claystone, upper part sandy, medium to dark gray, weathers yellow to yellowish brown, marine fossils common, 175-500 ft. thick; lower part sandstone, conglomeratic at base, indurated, yellowish gray to yellowish brown; 5-35 ft. thick. |
| | Boquillas | | Clayey, flaggy limestone and chalk, separated by thin marl layers; up to 700 ft. thick. |
| Lower Cretaceous | Buda | 93-95 | Limestone, grayish white; 65 ft. thick at Elephant Mountain. |
| | Del Rio Clay | 95-100 | Mostly claystone, some interbedded limestone and sandstone; claystone, soft, bluish to greenish gray, weathers yellow to light brown; 60-120 ft. thick. |
| | Santa Elena Limestone | +100 | Gray to brownish gray limestone, nodular in upper 40 ft.; about 740 ft. thick at Santa Elena Canyon, thinning northwards. |

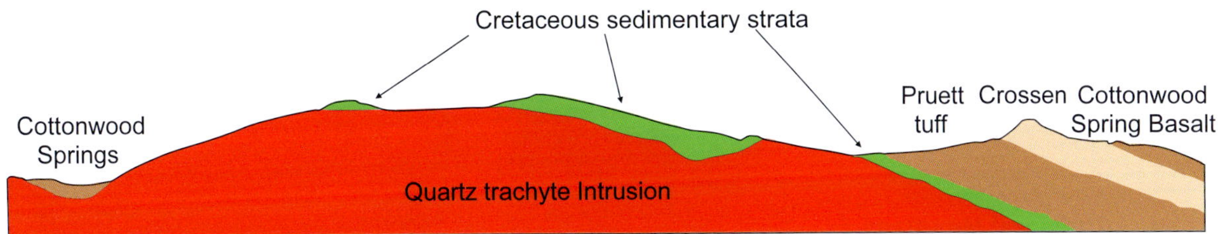


**Cienega Mountain**

**Above** Cienega Mountain (6,580 feet) at 1 o'clock from Cathedral Mountain Road at Mile 13.7 is a mushroom-shaped intrusion that has domed overlying Cretaceous limestones which can be seen as light-colored beds to the right of the dome. Cliffs 600 feet high face the upper part of the intrusion, a dome measuring about 3 miles by 2. The mountain is a trap-door laccolith, raised up from the left. The "door" is a sequence of light-colored Cretaceous limestone and sandstone beds dipping to the right, overlain by Pruett, Crossen and Cottonwood Spring Basalt strata. Some of the limestone was turned into marble by the heat of the intrusion.

intrusions, mostly trachytes, stand up above the lavas, including Ranger Peak at 2 o'clock and Twin Peaks at 3 o'clock. At Mile 3.5, the high rounded hill on the left is a remnant of Cottonwood Spring Basalt overlying Crossen lava. McIntyre Peak at 3 o'clock from Mile 4.4 is a very narrow and steep sided intrusion. An intrusive dike-like body of syenite outcrops at 10 o'clock.

At Mile 4.8, Highway 118 climbs up Big Hill out of the Alpine Basin onto a volcanic plateau which the highway traverses for the next 25 miles. The plateau is more than 30 million years old; deep valleys have developed in it between rough lava hills.

The volcanic succession on the plateau begins with a base of Pruett tuff (see table opposite). Although it averages 600 feet in thickness, it is a soft and easily eroded rock and seldom crops out except in road cuts and occasional hillside scars. The tuff usually contains calcite, probably from erosion of the underlying Cretaceous strata, and, in places, beds of non-marine limestone. For example, a half-mile beyond the picnic area at Mile 25.5, non-marine limestones crop out on the flanks of the hill on the right and 200 yards ahead, blocky light gray limestone overlain by volcanic rocks appears in road cuts on the right. This Pruett sequence of limestone and calcareous tuff beds is 300 feet thick with the upper 200 feet being nearly pure light gray to brown limestone. The limestone formed in saline lakes that developed from time to time on the Pruett landscape during a time when the climate was mild and wet.

Overlying the Pruett tuff is the widespread purplish-brown rhyolitic to trachytic Crossen lava which outcrops over a 400-square mile area from Alpine to Kokernot Mesa and from Mount Ord to Cienega Mountain. The unit may well continue under younger lava

**Mount Ord**

**Previous page** Mount Ord (6,700 feet) is the highest point in the Del Norte Mountains. The mountains form a 35-mile long ridge from Highway 90 south to the Santiago Mountains. The ridge is tilted up with its west face dipping towards Highway 118 at an angle of 10°.

The dip of the west face is at nearly the same angle as a bed of Crossen lava, 265 feet thick at the Mount Ord summit. In places the lava has been eroded away, exposing underlying soft Pruett tuff.

The lava breaks off at the crest of the ridge, creating a steep escarpment on the east, 600 feet high at Mount Ord. The escarpment is faulted along its base parallel to the ridge so it appears that the mountain range was uplifted after eruption of the Crossen lava, i.e. after about 36.8 Ma, by the third tectonic disruption mentioned in the Introduction.

Photographed from Mile 15.1.

## Cathedral Mountain

Cathedral Mountain (6,800 feet) is the most distinctive landmark on the volcanic plateau. You first see it from the northeast while climbing up to Mile High Road at Mile 10 and last see it from the southeast at Calamity Creek bridge. The mountain is an enormous block rising almost 2,000 feet above its surroundings, two miles long, one mile wide and weighing nearly 4 billion tons.

The upper bench of the mountain, seen in this photograph of the south flank, runs east-west and is capped by 545 feet of Petan Basalt above 450 feet of light-colored Tascotal Formation sedimentary strata mainly derived from volcanic rocks, called *volcaniclastic* strata by geologists.

Below the Tascotal strata is a bed of Mitchell Mesa Rhyolite, a welded tuff 85 feet thick. It caps two mesas and a butte 600 feet high and running three miles northwest from the main mountain, and just visible at the far left in this photograph. Under the welded tuff, creamy white Duff fine-grained rhyolitic tuff and volcaniclastic strata crop out sporadically along the mesa and butte.

The spire on the right, known as Haley's Peak, is a fault block thrust up by an intrusion that crops out on the lower slopes. Below the basalt on the spire, yellow Tascotal tuff and tuff breccia crop out in a nearly vertical cliff about 100 feet high.

The Duff, Tascotal and Petan Basalt strata are thicker on Cathedral Mountain and Goat Mountain than elsewhere, and were probably deposited in a basin that began sinking during the Duff period. The thick basalt slowed down erosion and accounts for the prominence of the mountains.

Photographed from Calamity Creek bridge at Mile 20.6.

## Elephant Mountain

Elephant Mountain (6,230 feet) dominates the landscape to the east of Kokernot Mesa. The mountain is capped by an enormous nepheline syenite sill, four miles long, two miles wide and 1,200 feet thick, weighing about 3 billion tons. The sill is above Cottonwood Spring Basalt on the west and below it on the east. The mountain was named for its shape, which resembles an elephant's back when viewed from some angles. The thin line of light gray tuff below the outcrops marks the approximate base of the intrusion.

The entrance to the 23,000-acre Elephant Mountain Wildlife Management Area (WMA) is on the left at Mile 26. Its principal wildlife species are javelina, pronghorn, quail, and dove. Bighorn sheep were introduced in 1987, although not native to the Trans-Pecos. The Area's web site is *www.tpwd.state.tx.us/wma/wmarea/elephant.htm.*

Photographed from Mile 32.4.

## Santiago Peak

Santiago Peak (6,521 feet) at 9 o'clock from its sign is one of the most striking landmarks in the Big Bend, rising 2,750 feet very steeply above Highway 118. The upper part is a nepheline syenite intrusion 1,250 feet thick and about three-quarters of a mile in diameter. Its lower boundary is near the light-colored rock debris about 40 per cent down the slope. The debris covers the boundary so it is not possible to say whether the intrusion is a plug or the remnant of a larger sill such as the ones capping Nine Point Mesa and Elephant Mountain. Its shape suggests that it is a plug.

The intrusion overlies 900 feet of Devil's Graveyard volcaniclastic sandstones, the only substantial body of volcanic-derived material east of TX 118. Boquillas limestone flags, 450 feet thick and also protected by the intrusion, stand up above the plain under the volcaniclastics.

The dark Black Mesa (4,610 feet) in front of Santiago Peak is a trachyte sill 3 miles long and 600 feet thick.

Photographed from Mile 32.4.

flows to the west. It averages about 200 feet in thickness on the plateau, the thickest flow, 265 feet, occurring on Mount Ord. The rock is well displayed in road cuts near the summit of Big Hill.

The Crossen lava was probably produced from multiple scattered fissures. One such fissure has been identified at Last Chance Mesa north of Alpine. There, a large dike that formed in the fissure when eruption ended can be seen joining a lava flow on the mesa. It is labeled "Crossen Dike" on the satellite photograph on page 22.

Above the Crossen lava, outcrops of the Sheep Canyon Basalt and Potato Hill Andesite members of the Pruett Formation can be seen in places, all overlain by multiple basalt flows and interbedded tuffs of the Cottonwood Spring Basalt. The highway mostly runs on these flows from Mile High Road until it comes down into the Calamity Creek valley near Elephant Mountain. The Cottonwood Spring lava is a dark gray to black basalt so most of the lava terrain is dark and rough with large dark boulders scattered down hillsides.

These basalt flows first appear just before Mile High Road at Mile 9.6, where the road crosses a saddle in which Cottonwood Spring basalt, broken up by a number of small faults, is exposed in road-cuts on the left and right. The saddle (5,403 feet), the highest point on Highway 118, is the divide between the Rio Grande and the

**Kokernot Mesa**

**Above** The elegant silhouette of Kokernot Mesa dominates the skyline as you leave the volcanic terrain of the southern Davis Mountains. The mesa rises 800 feet above the road on the right. Its cap rock is Crossen lava, 120 feet thick and jointed into columns. The column on the right is separated from the main part of the cap rock and seems ready to fall down the cliff. Light-colored Pruett tuff crops out below the cap rock.

Photographed from Mile 32.1.

Pecos River drainage basins. To the north of this point, creeks flow towards the Pecos River, to the south, towards the Rio Grande.

After crossing a beautiful grassy section of the plateau past the Border Patrol Inspection Station (see photograph on page 26), the road begins descending along the valley of Ash Creek from Calamity Creek Road at Mile 15.5 through nearly continuous road cuts in Cottonwood Spring Basalt. Interspersed in the lava flows are thin beds of tuff and at various points pink tuff can be seen below basalt beds in road cuts. The hot basalt baked the underlying tuff, turning black iron oxide ($Fe_2O_3$) into reddish hematite ($FeO$).

Looking ahead at Mile 20.0, the trace of a fault runs down Calamity Creek valley to the right of Elephant Mountain. The fault crosses the road about 200 yards before the bridge over Calamity Creek at Mile 20.5 and displaces basalt on the left down 525 feet against Crossen lava on the right. Weak and broken rock along the fault has helped Calamity Creek cut a valley along the fault trace. The creek is spring-fed along this stretch and always has water in it.

From the flat near Elephant Mountain at Mile 25.7, the road runs on Pruett tuff for the next six miles, very little of which is exposed. The Crossen lava capping the mesas on the right of the road rises gradually going south and ends with Kokernot Mesa (4,946 feet) (originally called Crossen Mesa) on the right at Mile 32.5.

## The Rio Grande Rift Zone

To produce this shaded relief map, the U.S. Geological Survey compiled the National Elevation Dataset of elevations every 10 meters (about 33 feet) in the United States and plotted it so that, by computer wizardry, it has been made to appear as a three-dimensional representation with the sun shining across the terrain from the west. The advantage of this method is that topographic features show up with great clarity, even better than in satellite photographs.

The Rio Grande Rift Zone is outlined in blue, solid lines where known, dashed lines where less certain. Notice how the zone remains quite narrow north of the Capitan Mountains but branches out into multiple segments south of that point.

The Salt Basin and Rio Grande rift segments come together at the Big Bend and continue into Mexico as the Sunken Block.

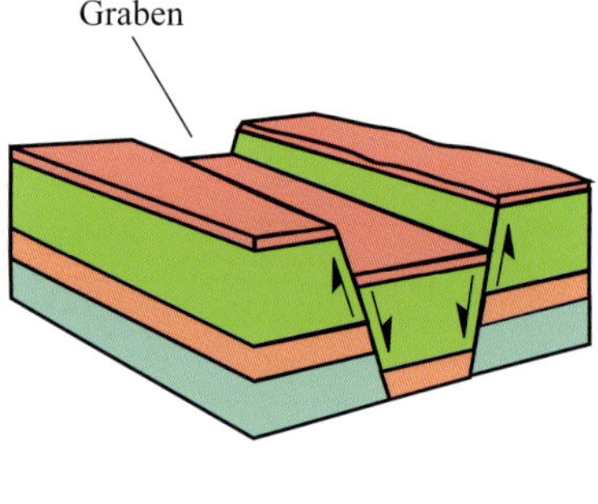


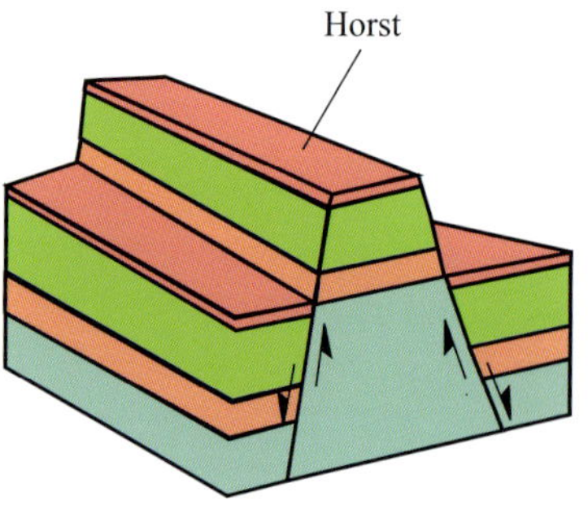

### Salt Basin Rift Segment

At about Mile 31.5, the highway crosses on to flaggy limestone of the Cretaceous Boquillas Formation, and two miles later, descends the 160-foot escarpment of the Walnut Draw Fault cutting through Boquillas and Buda limestones on the way down. The fault comes down Walnut Draw between Mitchell and Kokernot Mesas at about 4 o'clock and crosses the road diagonally. Strata are dropped down 1,100 feet between Kokernot and Mitchell Mesas so that the much younger Mitchell Mesa Rhyolite capping Mitchell Mesa is only 400 feet higher than the Crossen lava capping Kokernot Mesa, rather than the 1,600 feet it would otherwise be. At the highway, displacement on the fault is 750 feet to the south, decreasing to 300 feet near Black Mesa at 10 o'clock. The fault then disappears into alluvium.

The Walnut Draw Fault is on the northeastern edge of the Salt Basin rift segment discussed in the Introduction. The development of the Rio Grande Rift, named after the river which runs down it from Colorado to the Big Bend, approximately coincided with uplift of the center of the continent along a ridge in the Rocky Mountains from El Paso in Texas to Casper, Wyoming and on into Canada. Recent research shows that limited uplift began about 25 Ma, was quiescent for a time, and resumed more strongly at 15 Ma. Uplift is still ongoing.

Late Cretaceous sedimentary rocks such as the Aguja Formation show how much uplift has taken place. These rocks were deposited in lagoons at sea level but are now found at some 4,500 feet above sea level just south of Elephant Mountain, for example. Around Denver in Colorado, the uplift is over 5,000 feet. The Great Plains slope away from the ridge to the east at about five feet per mile; the top of Cretaceous strata at Dallas is 800 feet above sea level. The continent also slopes away from the ridge to the west.

Both the uplift and the rift are thought to have been produced by upwelling currents deep in the Earth's interior, probably resulting from subduction of the Farallon Plate, the subduction that generated the volcanic activity described earlier (page 17). Such currents sent heat and mass upwards creating magma chambers in the asthenosphere, the somewhat plastic zone 60 miles below the surface, uplifting the solid outermost shell of the Earth's crust, the lithosphere, and forming broad arches in which the crust thinned. Faulting then occurred along the uplifted crest into which magma oozed, leading to further uplift as the temperature increased and creating deep grabens.

The Rio Grande Rift began developing around 27 Ma, at first quite slowly. By 25 Ma, shallow basins filled by volcanic ash had developed along its length and for the next 10 million years it remained broad and shallow. At about 15 Ma, uplifting and block faulting intensified to create the rift as it appears today.

#### Graben Diagram

**Left** A graben develops when a block of the Earth's crust drops down below the level of surrounding strata along normal faults. A normal fault slopes inward with depth.

#### Horst Diagram

**Left** A horst is a block of strata left above its surroundings after they drop down along normal faults. Segments in the Rio Grande Rift often have horsts within grabens.

Grabens developed along faulting parallel to the rift as blocks sunk relative to the uplifting ridge. South of Socorro, New Mexico, the zone widened into a series of parallel rift segments, the easternmost of which is the Salt Basin segment, named after the Salt Basin at the Texas-New Mexico border.

From about 25 to 18 Ma, basalt lava poured out on the landscape after rising to the surface along the extension faults. This lava is called the Petan Basalt on the geological maps.

The main rift in Colorado and New Mexico is now more or less dormant, and consists of four major linked grabens up to 26,000 feet deep filled with lake and river sediments.

The Salt Basin rift segment may still be in an early stage of development. Its grabens are shallow; the deepest is a 4,000-feet deep graben at Valentine, north of Marfa. Recent faults are found along the length of the Salt Basin segment and earthquakes, which come from movement across faults, periodically occur in its vicinity, such as the 1994 earthquake near Alpine, and the largest earthquake recorded in Texas history, at Valentine in 1931.

## Green Valley

Having descended the Walnut Draw Fault escarpment, the highway enters the flat-bottomed Green Valley, 600 square miles of creosote and mesquite with occasional grass flats in low-lying areas and along Terlingua Creek. The area was given its name by early settlers who arrived during a rainy period when the grass grew high. The first rancher to write about Green Valley was Frank Collinson, a native-born Englishman whose memoirs, *Life in the Saddle*, make fascinating reading for students of West Texas history. In 1888, Collinson was given the opportunity to run a cattle herd near Straddlebug Mountain, 12 miles ahead, in return for a half-interest in the herd. The ranch's brand was said to resemble a straddlebug, the local name for the dung beetle. For seven years Collinson ran 7,500 cattle on 40,000 acres but by 1885 he had had enough and trailed his herd to the Pecos River. The herd only survived because the cowboys chopped up sotol plants to feed the cattle along the way. His reflections on the Big Bend were unflattering:

> "..the Big Bend is another Pharoah's Dream – a few good years are followed by more lean years that eat up all that the good years have made, and then some"

Nevertheless, in 1892, William Turney, an Alpine attorney, set up the O2 Ranch in the middle of the worst drought Collinson had ever seen. The ranch somehow survived and eventually took over most of Green Valley, including the Straddlebug Ranch.

After changing hands several times, the 268,578-acre ranch (420 square miles) was bought in 1941 for $805,000 by Lykes Brothers Inc., which still owns it. The family-owned firm is run by descendants of seven brothers who began in business shipping cattle and citrus fruit from Florida to Cuba. The family now owns 600,000 acres of land in Florida and Texas. The name of the O2 Ranch, incidentally, comes from the survey block in which the ranch lies.

The Green Valley basin stretches from the Walnut Draw Fault to Agua Fria Mountain, 25 miles ahead. It may have begun as a valley along Terlingua Creek which runs diagonally across it in a shallow channel, but then escarpments developed and as they receded left pediments or flats behind. The boundary is now 10 miles west of the creek, a beautiful volcanic escarpment from Bandera Mesa to Mitchell Mesa, 50 miles long and rising up to 1,000 feet above the creek. The escarpment is capped by Mitchell Mesa Rhyolite which is hard and durable in this area; it shows up as yellow-brown on the satellite photograph on page 14.

Terlingua Creek is the main drainage outlet for western Brewster County rising in Cartwright Mesa west of Green Valley and flowing south in a shallow channel through Green Valley to Hen Egg Mountain, descending at 30 feet per mile.

### Boat Mountain

**Left** Almost 400 feet of Pruett tuff is exposed on the flanks of Boat Mountain (4,345 feet), overlain by Cottonwood Spring Basalt lava, the most southerly exposure of this formation. The lava here consists of up to nine separate flows, 300 feet thick at maximum. To its right irregular low hills, also of basalt, end in a rough bluff typical of volcanic erosion.

Left of the highway, Green Valley laps up against fault blocks and intrusions such as Nine Point Mesa and where unimpeded stretches as far east as the Santiago Mountains. On the right of the highway, Green Valley is underlain by stream-borne strata of the Devil's Graveyard Formation. This formation is a combination of the Pruett and Duff Formations seen farther north. The intervening Cottonwood Spring Basalt and the Pruett lava members are missing south of Boat Mountain.

The Devil's Graveyard strata are 1,500 feet thick at Bandera Mesa, made up of 250 to 300 feet of limestone conglomerate at base, followed by a succession of floodplain, swamp and lake mudstones, and braided stream conglomerate and sandstone deposits. Fossilized crocodiles, sharks and large trees show that the climate was warm and humid during the formation of the Devil's Graveyard Formation, 47 to 32 Ma. The lower conglomerates contain only a few volcanic rock fragments but the proportion rises steadily up the succession. Initial volcanic material probably came from Mexico, but by the end of deposition the Chinati volcanic center was active and supplied volcanic ash to the upper strata.

### Santiago Peak

**Right** Santiago Peak dominates the east side of Green Valley.

The cockscomb ridge immediately to the right of Santiago Peak, not shown in the photograph, is Y E Mesa (5,385 feet), a syenite sill two miles in diameter and about 1,400 feet thick. The name comes from a cattle brand used locally in the early twentieth century. To its right the dome-shaped Red Mountain (4,100 feet) is also an intrusion.

Photographed from Mile 41.1.

### McKinney Peak

**Left M**assive swaybacked McKinney Mountain (5,006 feet) is 5 miles east of the main body of the Mitchell Mesa Rhyolite. Only a narrow band of the welded tuff remains on top of the mountain but it is 150 feet thick rather than the normal 40 to 75 feet, so this extra thickness probably explains why it has survived so far in front of the main escarpment. The peak is named for W. Q. McKinney, who built a rock house nearby in 1886.

On Turney Peak (4,854 feet) to the right front, the welded tuff has been eroded off completely, leaving the underlying Duff Formation strata unprotected. Its rounded top shows that it is being eroded rapidly. Its name comes from William Turney, who founded the O2 ranch in 1888.

Photographed from Mile 34.7.

The Devil's Graveyard Formation at one time extended across Green Valley to Santiago Peak as the 900 feet of Devil's Graveyard volcaniclastic strata on the peak prove. No trace of these rocks have been found east of this point but it is unlikely that they did not extend beyond Santiago Peak as the top of the Devil's Graveyard strata is well above the summits of the Santiago Mountains.

The Green Valley area has been lightly folded into low synclines and anticlines, now much broken up by faulting. One such anticline runs across the basin towards its northwest corner. Strata are 500 feet higher at the roadside at Mile 43.5 than at Santiago Peak. They then dip 1,300 feet down the other flank of the anticline to Terlingua Creek. This is another example of folding from compression in the third tectonic disruption of the Introduction. The Mitchell Mesa Rhyolite has been affected by this folding, so it took place after 32 Ma when the rhyolite erupted.

The highway runs through Green Valley on alluvium and Boquillas Formation flagstones, the volcanic and volcaniclastic rocks having been eroded away on the east side of the basin. Boquillas limestones break up easily and seldom form more than low hills. Flaggy fragments litter the roadside and show up in road cuts.

## The Sunken Block

At Mile 43.2, the highway climbs up the low escarpment of the Chalk Draw Fault. The fault, although it only has a 300-foot downthrow here today, is part of a major structural line running from the Chinati Mountains to Del Rio. The fault has been active several times in geological history. In the early Permian period, for example, the fault displaced strata 10,000 feet down to the north.

Looking to the left (east) at bottom of the rise, the fault escarpment runs down the narrow Chalk Draw valley. The horst above the escarpment (see diagram of a horst on page 38) is capped by Boquillas and Buda limestones with the brownish slope below underlain by Del Rio shales. At base of the escarpment is a 100-foot cliff of Santa Elena Limestone.

Beginning at the Chalk Draw Fault, strata begin to sink down into the broad rift valley known to geologists as the Sunken Block. The 30-mile wide rift valley is divided into eastern and western grabens by a central horst that includes the Christmas and Chisos Mountains (igneous blocks that resisted subsidence) and continues into Mexico along a series of anticlines for 25 miles or so.

### The Chisos Mountains

**Above** The Chisos Mountains, seen here on the skyline in the early morning light, are prominent on the skyline for several miles from the Chalk Draw Fault escarpment.

The Corazones Peaks in mid-picture are intrusions northeast of the Christmas Mountains. The east peak (5,045 feet) on the left is part of a trachyte intrusion. The west peak (5,319 feet) stands alone, a rhyolite intrusion.

Casa Grande (7,325 feet), in the Chisos Mountains, is just to the left of the west peak.

In the map on page 46 (an enlargement of the one on page 38), the Sunken Block is shown as a continuation of the Salt Basin segment that joins the Rio Grande segment at the Big Bend. The Chalk Draw Fault is both the northern boundary of the Sunken Block and the central horst.

Faulting on the west begins as a series of step faults, shown as black dashed lines on the map, and then becomes a single fault along the Christmas and Chisos Mountains before returning to step faulting near the river.

On the east, a single fault, a continuation of the Chalk Draw Fault, separates the horst from the grabens to its east. Strata in the horst are a maximum of 1,500 feet above strata in the grabens on either side.

Strata on both sides of the Sunken Block are about 3,000 feet lower at the river than are strata outside the block. The Block bifurcates in Mexico and continues a substantial distance to the south-southeast, as shown by the dashed lines on the map opposite. Grabens in the Block deepen in Mexico, to 6,000 feet along the Sierra del Carmen and to 4,500 feet along the Terlingua Fault.

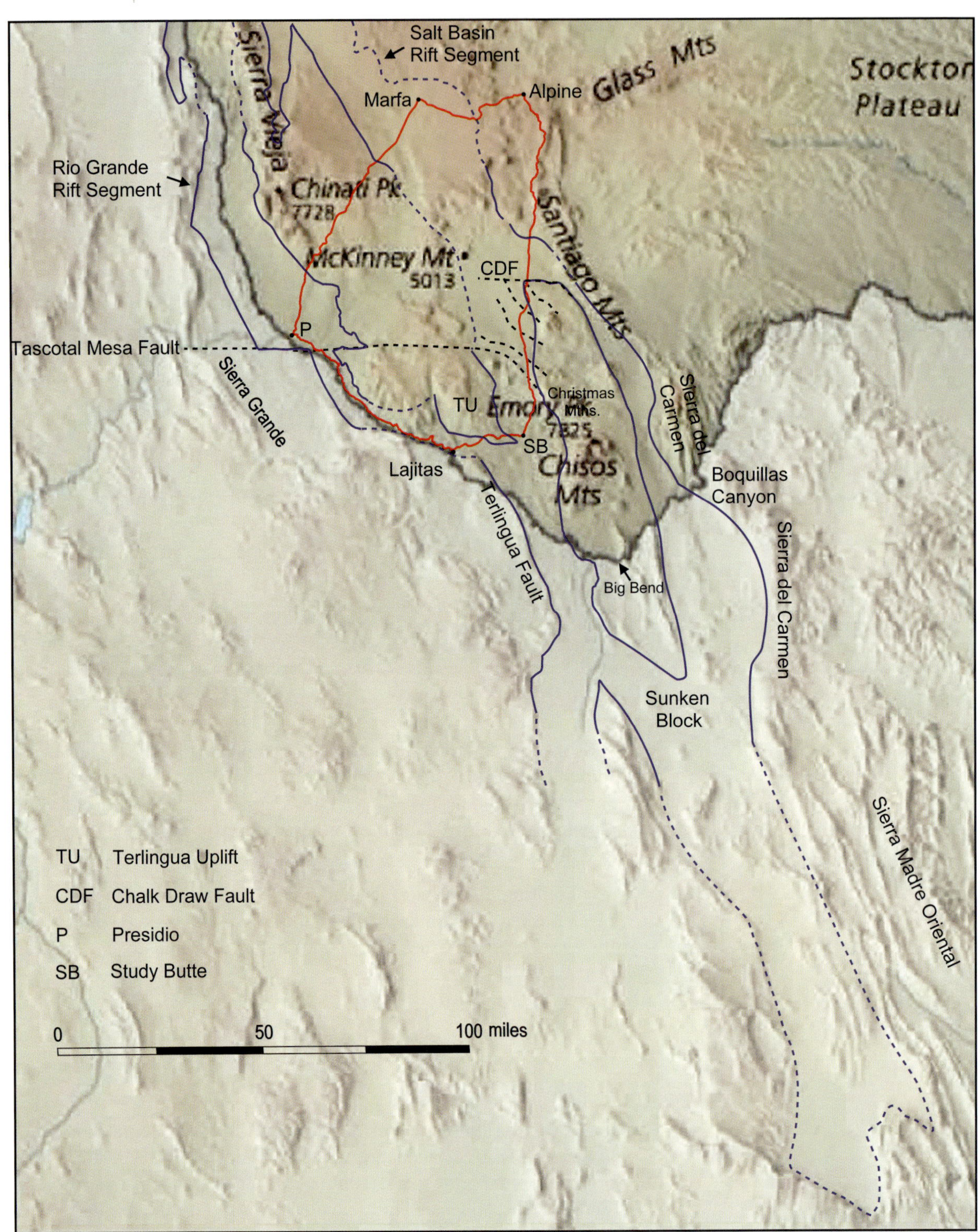


**The Rio Grande and Salt Basin Rift Segments and the Sunken Block**

The Sunken Block parallels the numerous south-southeast ridges in West Texas and Mexico. These ridges result from the Earth's crust coming into compression from the southwest in the first of the tectonic disruptions mentioned in Chapter 1. Compression resulting from the subduction of the Farallon Plate had begun on the west coast 150 Ma and migrated east at about one inch per year, producing folding, buckling and thickening of the Earth's crust across the western continent.

The final mountain-building episode, the Laramide Orogeny, created mountain ranges from British Columbia in Canada to Sierra Madre Oriental, including the Santiago and Sierra del Carmen Mountains. It also created the numerous tightly-folded and faulted anticlines that appear on the map opposite as ridges as well as the Terlingua Uplift (see page 63). The orogeny took place in two pulses, one from 70 to 66 Ma, the other from 61 to 59 Ma according to sedimentary evidence in Big Bend National Park.

The first of the step faults on the east is encountered on top of the Chalk Draw Fault rise, where the road turns left along a northwest-southeast fault with a displacement of 600 feet down to the southwest. A windmill and mesquite trees near the road on the left indicate water present along the fault.

The faults form low escarpments facing south on which Boquillas limestones crop out. At Mile 50.2, the road climbs up the escarpment of the next fault past Boquillas strata exposed in road cuts. The formation here consists of 12-inch limestone beds interspersed with dark marl beds 8-10 inches thick. Marl is a mixture of clay and limestone. The fault drops down strata 500 feet to the southwest and is approximately in line with Cartwright Canyon in Nine Point Mesa ahead.

Two more northwest-southeast faults cross the highway at Mile 57.5 just before and just after Agua Fria Road with a combined down-throw of 800 feet to the southwest.

### The Sunken Block

The Rio Grande and Salt Basin Rifts are shown coming together at the Big Bend into the Sunken Block.

Rift segment outlines are shown as blue lines, solid where known, dashed where less certain.

The extension of the Sunken Block into Mexico has been mapped from aerial photographs and is speculative.

## The Devils Graveyard

Badlands in the Devil's Graveyard photographed from Mile 51.0. A bentonite quarrying operation can be seen in mid-picture. The clay mineral bentonite occurs as irregular lenses in Devil's Graveyard Formation tuffs near the base of the formation, the weathered product of volcanic ash.

After being quarried, the mineral is ground into powder and bagged. Bentonite is used as a lubricant in drilling mud and as a binder in animal feed pellets.

The formation was named after this location.

## Red Bluff

**Above** Red Bluff (4,018 feet), left of the road at Mile 53.5, is a narrow crescent-shaped wall rising 600 feet above the surrounding flat and 350 feet above a laccolithic intrusion behind it. The wall was injected into a fault created when the laccolith was intruded and was later uncovered by erosion. The two Corazones peaks are behind and to the right of Red Bluff. These are the first of the many intrusions seen around the Christmas Mountains.

## Agua Fria Mountain

**Right** Agua Fria Mountain (4,828 feet) is seven miles away at 2 o'clock from the Agua Fria Road intersection at Mile 57.5. It is a dissected rhyolite body two miles in diameter that towers 1,400 feet above the surrounding countryside. Huge, irregular columns of reddish weathered rock form cliffs around its periphery. Panther Mountain (4,331 feet) is the craggy peak to its left, a rhyolite trapdoor laccolith like Cienega Mountain (page 29), surrounded by sheer cliffs 600 feet high. The mountain retains some of its sedimentary cover; Santa Elena Limestone strata dip steeply off its southern side.

### Igneous Intrusions around the Christmas Mountains

From Mile 63.5 onwards, the highway threads its way between an abundance of igneous intrusions rising out of the Cretaceous strata. Many rise in steep rocky crags 700 to 1,500 feet high. Most are rhyolite, quartz trachyte or trachyte. Such rocks consist mainly of light-colored quartz and feldspar with light gray fresh surfaces. Weathered surfaces are typically dark red or orange red from iron oxide staining.

The intrusions along Highway 118 are part of a cluster around the Christmas Mountains (geological map on page 20). The Christmas Mountains have a long history of igneous activity, beginning at 47 Ma, when a gabbro laccolith was intruded west of Little Christmas Mountain. Volcanic eruptions occurred on top of the mountains at about 45 Ma, and a rhyolite laccolith formed a dome there at 42 Ma. Most intrusions along Highway 118 fall in the age range 44-40 Ma and are very similar in composition to the Christmas Mountains lavas. This history points to a long-lived magma pool under the mountains, the source of the lavas, tuffs and intrusions.

## Nine Point Mesa

**Above** Nine Point Mesa (5,502 feet), photographed from Mile 63.0, is a massive presence six miles east and 2,000 feet above Highway 118. Roughly circular, six to seven miles across, it is capped by two sills of quartz trachyte 1,000 feet thick, which form 200-ft igneous cliffs set on a pediment of Pen and Boquillas sedimentary rocks. It is among the oldest intrusions in the area, one of its sills being about 46 m.y. old. Two deep canyons following northwest-southeast faults dissect the intrusion. The northeastern end of one, Cartwright Canyon, can be seen from the road at Mile 60.5.

The small double hill on the right of the photograph is Camel's Hump (3,662 feet), a rhyolite plug intruded into Boquillas limestone. Though it was probably named for its shape, Camel's Hump also lay on the route of a United States Army camel expedition through this area in 1859, a War Department experiment to see how camels could be used in the Southwest. The answer was that they did not work very well. Their feet were cut up by the sharp rocks so prevalent in the Big Bend and the experiment was abandoned on the outbreak of the Civil War. Stray camels were sighted for several years in West Texas.

## Packsaddle Mountain

**Right** Packsaddle Mountain (4,658 feet) rises 1,200 feet above the road 1½ miles west of the sign for the mountain at Mile 62.9. It is a rhyolite plug in which the intrusion has broken through Santa Elena Limestone. Light-colored beds of limestone remain on its flanks, particularly on the northeast side, tilted up by the intrusion.

Near the road, two dikes run parallel to the road, creating iguana-like ridges.

## Hen Egg Mountain

**Above** Hen Egg Mountain (5,002 feet), 3½ miles west of the sign for the peak at Mile 64.1 is a rhyolite plug, almost perfectly symmetrical, 40 million years old, and one of a group of four intrusions, most likely from a common source. The upper 1,500 feet of the mountain is above the level of surrounding Devil's Graveyard strata at the time of the intrusion, and so must have broken through into the open air. Another peak of the group is behind Hen Egg Mountain to the right.

## Christmas Mountains

**Right** At the bridge over Dark Creek Canyon at Mile 70.0 the Christmas Mountains (5,728 feet) stand 2,500 feet above the road at 9 o'clock, a magnificent, craggy, eroded dome. Light-colored beds of Santa Elena Limestone dip to the right off the dome elevated as much as 5,000 feet above their normal position, probably by an unexposed intrusion. Volcanic breccia, produced by two small calderas on the dome, caps the highest summits. A fault escarpment 1,200 feet high faces the road. The fault most likely developed during the doming and was reactivated during extension of the Salt Basin rift segment.

Photographed from Mile 69.9.

The Highway 118 intrusions are quite small, generally a mile or less in diameter, and can be thought of as blobs rising up from a magma reservoir below and solidifying before they reached the surface. The intrusions are fine-grained, indicating that they were intruded at no great depth; shallow intrusions cool more rapidly and so have smaller crystals. They seem to be concentrated at the boundary between Upper Cretaceous claystones and overlying Devil's Graveyard tuffs which have now been eroded away. Some of the narrower intrusions may have been feeders for lava flows in the Devils Graveyard or younger strata that have also been eroded away.

North County Road on the right at Mile 65.7 is part of the old Terlingua-Alpine road but is only open to the public now for 3.5 miles. The graveled road descends 500 feet over steep grades into the valley of Terlingua Creek and provides excellent views up the valley to Hen Egg and Agua Fria Mountains.

Just beyond North County Road, the Tascotal Mesa Fault crosses the highway. This fault runs from Highway 118 across the Bofecillos Volcanic Field and the southern boundary of the Presidio Basin to

## Paisano Peak

**Above** Paisano Peak (5,451 feet), not to be confused with the Paisano Peak west of Alpine (page 172), is a rhyolite trapdoor laccolith, 45 million years old. The 1,200-feet escarpment facing the road is the trapdoor opening. On the opposite side of the mountain, remnants of Buda and Del Rio strata slope away from the summit. Fluorspar was mined on the west side of the peak from 1971 to 1980. The main source of the element fluorine, fluorspar is found in limestone near contacts with rhyolite intrusions in the Christmas Mountains and in Mexico.

Photographed from Mile 70.9.

## Pen Formation Claystone Badlands

**Right** About Mile 73.5, Cretaceous Pen Formation claystones, underlying the Aguja claystones, begin cropping out on the right. These soft strata have been preserved in the low-lying terrain under flat-lying intrusions and crop out in badlands, eroded into irregular purplish or yellowish gray mounds and pedestals.

An eroded laccolith capping the Pen strata forms a continuous ragged ridge above the road for a little over two miles.

Photographed from Mile 74.3.

Sierra Grande, where it coincides with a kink in the mountain, a total of nearly 90 miles. From this point west, the fault acts as a transfer zone in the Rio Grande Rift and is a transverse fault – strata moved horizontally along it (see page 82). To the east of the highway it becomes a complex zone of several normal faults with a 710-foot down-throw to the south and joins a fault running along the side of the Christmas Mountains ahead.

From about North County Road onwards, Cretaceous Aguja Formation claystones begin showing up at roadside. These soft strata, containing gypsum crystals that glitter in the sunlight, have been eroded into badlands of irregular mounds and pedestals. The Aguja strata were deposited in heavily forested coastal swamps and contain lignite beds in places, one of which, six miles north of Study Butte, was quarried by the Chisos Mining Company at Terlingua for its furnaces.

The company at first used local timber but as the Terlingua area rapidly became treeless, costly wood had to be imported from as far away as Alpine. Once this lignite source was found, the company set up equipment to blow steam and air through red-hot lignite, generating producer gas, a combustible mixture of carbon monoxide, hydrogen, carbon dioxide, methane, and nitrogen, to fire the furnaces.

### Little Christmas Mountain

Little Christmas Mountain (4,828 feet) is an eroded stack of trachyte and rhyolite lavas and ash flow tuffs which had erupted from the Christmas Mountains volcanoes.

Photographed from Mile 74.9.

## Willow Mountain

**Above** Willow Mountain (3,830 feet), a reddish brown quartz trachyte laccolith, 40 million years old, looms above the road 2½ miles north of Study Butte. Its south face provides the most spectacular display in the Big Bend region. Closely spaced, nearly vertical columns run down the cliffs from top to bottom. Fallen columns litter the ground below. Stands of native desert willows grow around Willow Spring at the base of the mountain on the south and it was probably named for them.

Photographed from Mile 75.3.

## Bee Mountain

**Right** Bee Mountain (3,452 feet) is at roadside on the right just before Study Butte, a reddish, craggy trachyte dome rising above the road in sheer cliffs up to 850 feet high. The fresh rock surface exposed in a road cut is a light greenish gray.

The intrusion sits on beds of Pen Formation claystones which form the light gray layer near ground level.

Photographed from Mile 75.6.

### Study Butte

The junction of TX 118 and Farm Road 170 in Study Butte is at Mile 76.6 on the flat just beyond Bee Mountain. Turn right at the junction to go to Lajitas.

The town, like Terlingua to its west, owes its existence to cinnabar mining. In 1905, Will Study (pronounced "stoody") developed the Big Bend Quicksilver Mine on the left flank of a butte, Study Butte, one mile south of TX 118-FR 170 junction. The butte is capped by a tongue-shaped trachyte intrusion, 36 million years old, intruded into Pen Formation strata and the source of mercury-bearing fluids. High-grade ore mostly occurred near contacts with sedimentary rocks or where fractures were concentrated.

The mine operated from 1905 to 1946, and again from 1970 to 1972, employing up to fifty people. By the time it closed, it was the last functioning cinnabar mine in the Terlingua mining district but the town by then had taken on a new lease on life as a tourist point at the west entrance to Big Bend National Park.

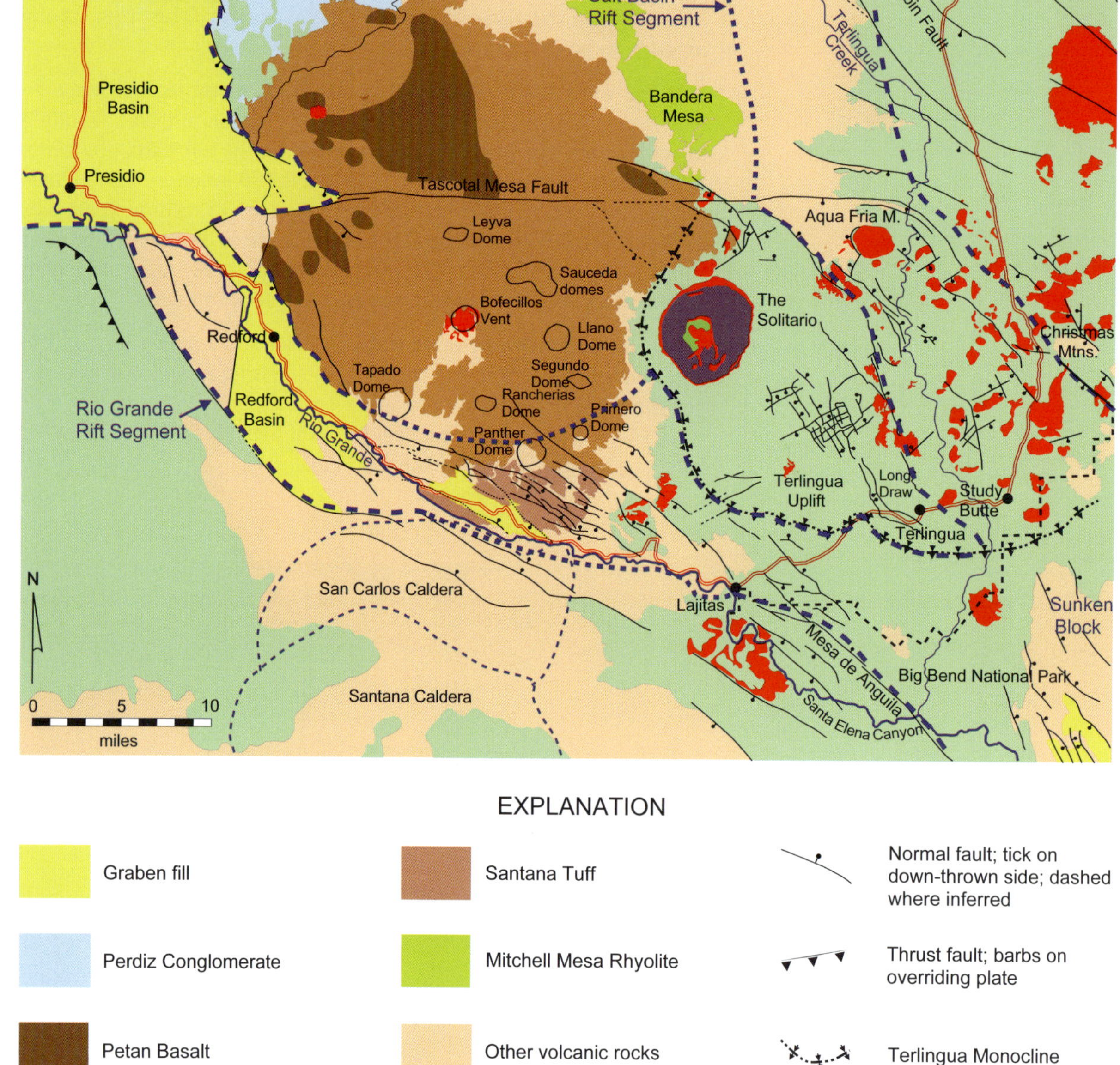


EXPLANATION

| | | |
|---|---|---|
| Graben fill | Santana Tuff | Normal fault; tick on down-thrown side; dashed where inferred |
| Perdiz Conglomerate | Mitchell Mesa Rhyolite | Thrust fault; barbs on overriding plate |
| Petan Basalt | Other volcanic rocks | Terlingua Monocline |
| Igneous intrusions | Cretaceous strata | Caldera boundary |
| Volcanic rocks of the Bofecillos Volcanic Center | Early Paleozoic strata | Rift segment boundaries dotted where inferred |

**Geology: Study Butte to Presidio**

# 3 Study Butte to Presidio

For the second section of the River Road loop turn right at the junction of TX 118 and Farm Road 170 in Study Butte. Mileages along the route are measured from the junction. FR 170 is not a road for fast driving, although perfectly safe when driven carefully. It has many sharp turns and several steep hills. The trip, 66.8 miles long, ends at the junction of FR 170 and US 67 on the west side of Presidio and takes about two hours, including time for photographs and other stops; no gasoline or water is available along the route. Strata along the route are described in the table on page 66.

## The Geological Map

The map opposite shows the geology along the Study Butte-Presidio road. The route begins at the point where the Salt Basin rift segment joins the Rio Grande rift segment in the Sunken Block. After climbing over the Terlingua Uplift, the road follows the the narrow Rio Grande valley between the Bofecillos Mountains and the mountains of the Santana and San Carlos Calderas. In the final seven miles, the valley opens up in the Presidio Basin.

Study Butte is in the Castolon Basin, a low-lying area between the Mesa de Anguila, the volcanic mountains to the east and the Terlingua Uplift. The basin, drained by Terlingua Creek which crosses FR 170 about two miles to the west, slopes imperceptibly down from 2,600 feet elevation at Study Butte to 2,150 feet overlooking the Rio Grande at Santa Elena Canyon. Sedimentary rocks of the Upper Cretaceous Aguja, Pen and Boquillas Formations have been mostly preserved from erosion in the basin although under attack along Terlingua Creek.

The Terlingua Uplift, four miles to the west of Study Butte is a block popped up by compression from the west-southwest during the Laramide Orogeny, 70 million years ago or so. Strata in the Uplift, which runs northwest from Terlingua to the Solitario, are up to 1,500 feet above strata to the south. The physical uplift is never more than about 900 feet above the basin, however, because the younger Cretaceous strata have been eroded off the uplift but remain in the basin. The boundary of the uplift on the west and south is the Terlingua Monocline, which brings down strata 500 feet at the Solitario, 800 feet where it turns east, and to nearly 1,500 feet south of Terlingua. It fades away in Big Bend National Park.

## The Panoramic View from above Terlingua Creek

In this wonderful panoramic view from the west bank of Terlingua Creek, Maverick Mountain (3,496 feet) is the dark brown intrusion in left middle distance. The Chisos Mountains are behind it on the skyline, with the sharp notch of Casa Grande near the right of page 64. The yellowish inclined hill in front is Study Butte (2,830 feet).

The high light-colored summit below the clouds is Emory Peak (7,825 feet), the highest point in the Chisos and the third highest summit in Texas. The level plateau to the right of the Chisos is the South Rim (7,400 feet), a favorite hiking destination.

Below the Chisos Mountains, the long low brown cliff is Burro Mesa, capped by a rhyolite lava.

The stream in the near foreground is Terlingua Creek, the major tributary of the Rio Grande between Alamito Creek, 60 miles to the west, and Maravillas Creek, 50 miles to the east. The creek is the main drainage outlet for western Brewster County, flowing

parallel to Highway 118 from the Chalk Draw Fault to Study Butte and draining 1,315 square miles. It rises in Cartwright Mesa west of Green Valley and flows in a shallow channel through Green Valley to Hen Egg Mountain, descending at 30 feet per mile. It has created a 180 foot deep valley from the vicinity of Hen Egg Mountain, 12 miles to the north, to its mouth at the Rio Grande, 10 miles to the south.

The yellow banks on either side are Upper Cretaceous sedimentary rocks, mainly mudstones and siltstones of the Pen and Aguja Formations. They were deposited near the end of the Cretaceous period, around 88 million years ago when the Big Bend was in the process of rising above the Cretaceous ocean and mud and silt were coming into shallow bays along the coastline of the seaway. At roadside at Mile 1.0, bluffs of the underlying finely bedded cream-colored Boquillas limestone can be seen in the banks of the creek.

Photographed from Mile 1.5.

## Strata from Study Butte to Big Hill

| Period/Epoch | Formation | Age m.y. | Description |
|---|---|---|---|
| Oligocene/ Eocene | Tule Mountain Trachyandesite | ~33 | Porphyritic trachyte lava, a single lava flow as much as 360 ft. thick in south-central and southeastern Bofecillos Mountains. |
| | Mule Ear Spring Tuff | 33.1 | Densely to poorly welded ash-flow tuff, a single ash flow up to 30 ft. thick; source Sierra Quemada (in Chisos Mountains) or to south in Chihuahua. |
| | Bee Mountain Basalt | 34.5 | Basalt lava; several flows in southeastern Bofecillos Mountains, probably from a series of broadly related basalt eruptions. |
| | Chisos | 47-34 | Tuffaceous sediment and conglomerate, massive to bedded, pumiceous sandstone, siltstone and conglomerate; up to 800 ft. thick, thinning out against Terlingua Monocline and the Solitario; occurs throughout southern Bofecillos Mountains and includes thin lenses between Bee Mountain Basalt, Mule Ear Spring Tuff, Tule Mountain Trachyandesite. |
| | Alamo Creek Basalt | 47 | Basalt lava forming single flow on northeast flank of North Lajitas Mesa. |
| Upper Cretaceous | Aguja | ~70 | Claystone, sandstone and lignite. Upper part continental deposits up to 880 ft. thick; sandstone, argillaceous, various shades of yellow and brown; claystone, calcareous in part, greenish gray to yellowish brown and purple; a few lignite beds; vertebrate fossils, petrified wood common. |
| | Pen | | Lower part 3 units; first transitional from upper unit; middle unit, silty to sandy claystone, medium to dark gray, weathers yellow to yellowish brown, marine fossils common; 175-500 ft. thick; lower unit sandstone, conglomeratic at base, indurated, yellowish gray to yelowish brown; 5-35 ft. thick. |
| | Boquillas | | Mostly claystone; upper part sandy; middle part, yellow scattered sandy beds; lower 50 ft. calcareous claystone with inch-thick chalk beds, light bluish gray, weathers yellow to yellowish gray; marine fossils throughout; 219-700 ft. thick.<br><br>*San Vicente Member*: limestone flags, interbedded with gray to yellowish gray platy marl and soft gray marl; marine fossils abundant; 130-400 ft. thick.<br><br>*Ernst Member*: limestone, siltstone and claystone; limestone, silty, flaggy, beds mostly 2-5 ins., some up to 18 ins.; bluish gray, weathers light yellowish gray to light brownish yellow, blocky from joints; marine fossils abundant; 450 ft. thick. |
| Lower Cretaceous | Buda | 93-95 | Limestone, grayish white; 25 ft. thick at Santa Elena Canyon. |
| | Del Rio Clay | 95-100 | Mostly claystone, some interbedded limestone and sandstone; claystone, soft, bluish to greenish gray, weathers yellow to light brown; 30 ft. thick at Santa Elena Canyon. |
| | Santa Elena Limestone | +100 | Fine grained to microgranular, massive, beds up to 10 ft. thick, light gray to white; weathers dark gray and shades of brown; forms cliffs; about 740 ft. thick at Santa Elena Canyon. |

On the east, the uplift is bordered by the Yellow Hill Fault along the west bank of Terlingua Creek. The fault drops strata down about 1,000 feet near Pinks Peak. It dies away into the Terlingua Monocline at Cuesta Blanca, south of the road at Mile 2.4, and into the higher terrain north of the Solitario. Several intrusions are buried in the uplift including one west of Pinks Peak which has domed strata above it and created the great concentration of faulting shown on the map.

The Solitario, a spectacular eroded dome near the north of the Terlingua Uplift, is an extinct volcano that erupted in three pulses between 36 and 35 Ma. In the first pulse, sills and dikes were intruded into Cretaceous limestones. In the second pulse, several hundred thousand years later, one of the world's largest laccoliths was emplaced, doming the rocks above. The intrusion was accompanied by an enormous ash eruption that emptied the magma chamber. The chamber's roof collapsed, forming a caldera with walls 1,000 feet high. The walls were fractured and over time chunks from them fell into the caldera, partially filling it with debris. The third igneous pulse intruded dikes and small laccoliths into ash-flow tuff and debris within the caldera.

Today the Solitario has a high rim of tilted resistant limestones (photograph on page 90). Several ash-flow vents are preserved including a 1,000-foot circular one on the west of the volcano filled with ash-flow tuff and rock torn from its walls. Early Paleozoic rocks of the Cambrian, Ordovician and Pennsylvanian periods crop out in the central basin along with the laccoliths and dikes intruded in the third igneous pulse.

A faulted monocline on the northern side of Mesa de Anguila provides the southern boundary of the Castolon Graben. The faulting, which took place with development of the Rio Grande rift segment, brings strata 3,000 feet down into the graben at the mouth of Santa Elena Canyon.

The Mesa de Anguila like the Terlingua Uplift is also thought to be a block popped up by compression from the west-southwest during the Laramide Orogeny. Another later period of compression appears to have caused local folding. An anticline in the west end of the Mesa de Anguila includes a basalt intrusive sill which has been dated at 33.7 Ma, long after the Laramide Orogeny. This later compression may also have caused the folding in Green Valley and perhaps along Mount Ord near Alpine described in Chapter 2.

The Castolon Basin narrows towards Lajitas. There the landscape changes dramatically as the road leaves the wide basin framed by light colored limestones to delve into a deep volcanic valley, 30 miles long, the narrowest part of the Rio Grande rift segment. Strata are stepped down along this valley by a series of west-northwest faults, as shown on the geological map on page 62.

In the Presidio Basin, the rift segment widens out into an enormous graben, 36 miles long and 18 miles wide, 10,000 feet deep on the west and at least 4,000 feet deep on the east. The graben is filled with sand and gravel, about 2,500 feet thick at Presidio.

### The Satellite View

In the photograph above, the boundaries of the rift segments are outlined with yellow lines, dashed where fairly certain, dotted where less certain. Leaving Study Butte, the road crosses Terlingua Creek at Mile 1.2 and bypasses the dark splotch of Cigar Mountain, a basalt intrusion. Several other dark intrusions can be seen on either side of Terlingua Creek.

The Terlingua Uplift runs from just south of FR 170 at Terlingua to beyond the Solitario. Strata in the uplift were arched by compression from the west-southwest between 68 to 50 Ma. Compression created a series of local east-northeast trending grabens and fractures, some of which can be seen on the photograph.

Igneous intrusions came later and introduced cinnabar, the ore of mercury, along faults and fractures parallel to the grabens. Intrusions that have been dated include Sawmill Mountain, which is 44 million years old (photograph on page 70).

As the Rio Grand rift segment began to form about 24 Ma, it wrenched strata south of the Tascotal Mesa Fault to the west in relation to strata to the north of the fault. This wrenching created another set of grabens trending north-northwest, the largest being Long Draw Graben which the highway drops into at Mile 4.9.

Erosion later stripped off strata of the arched center of the uplift down to the Santa Elena Limestone, the main lower Cretaceous formation in this area, except in the grabens where Buda, Del Rio, Boquillas and Pen strata still remain. The Santa Elena unit, which also caps Mesa de Anguila, appears in the photograph as a yellowish brown block. Younger Boquillas Formation strata crop out around the uplift and appear on the photograph as light yellow areas.

The Santana Caldera in Mexico is beautifully delineated with the great dome of Sierra Rica in its midst. The Sierra el Mulato, which dominates the scenery south of the river along the route, is also prominent.

The Presidio Basin is silvery gray in Texas but creamy yellow in Mexico. South of Redford the dark gray Redford Basin straddles the Rio Grande, the latter identified by its border of green irrigated fields.

A larger-scale satellite photograph of the route from Lajitas to Presidio is on page 87. The river valley and the Bofecillos Mountains are described in more detail there.

### The Terlingua Mining Area

Just after South County Road at Mile 3.8, part of the old Terlingua-Marathon road, FR 170 climbs 120 feet up the scarp of the Yellow Hill Fault on to the Terlingua Uplift and at Mile 4.5 passes the exit to the Terlingua Ghost Town on the right.

The ghost town belongs to a period of cinnabar mining in the area. This bright red mineral (mercury sulfide HgS) was used for wall and war paint for many centuries before being found to be a source of mercury near the end of the nineteenth century. Mercury has been used in thermometers, invented by Fahrenheit in 1720, as an amalgam in dentistry, as an amalgam used in extracting gold and silver from their ores, and to manufacture mercury fulminate, used in explosives and military cartridge fuses.

At Terlingua, cinnabar was found along the contact between Santa Elena Limestone and Del Rio Formation claystone, from calcite veins in the Boquillas and Buda Formations, and from breccia pipes.

## Looking North across the Terlingua Uplift

**Above** Looking across the Terlingua Uplift from just before the Ghost Town exit, Sawmill Mountain ((3,750 feet) is the dark hill on the left, a porphyritic trachyte plug 44 million years old. Peaking round its shoulder is Agua Fria Mountain.

Hen Egg Mountain (5,005 feet), a rhyolite plug, is 10 miles away at right center. It rises 2,000 feet above Terlingua Creek in front. Panther Mountain (4,331 feet), a similar plug, is just to its right. The small dark pyramid in front of Hen Egg Mountain is a basalt intrusion like Cigar Mountain.

## Santa Elena Canyon

**Right** The spectacular cleft of Santa Elena Canyon in the Mesa de Anguila cliff face, 11 miles away, photographed from the Terlingua Uplift at Mile 4.0.

The Rio Grande crosses the Mesa from the west, having cut down 1,500 feet in the last two million years. This is a good example of the superimposition of a river which had set its course when the terrain surface was a good deal higher than the mesa, and was able to cut down quickly enough to maintain its course.

The upper limestone bed in the cliff face is the Santa Elena Limestone, 740 feet thick. It also crops out on the Terlingua Upllift and can be seen above the road on the Reed Plateau in Long Draw at Mile 6.0. Below it, the darker slope is formed by the marl and limestone Sue Peaks Formation, 265 feet thick, followed by another massive limestone, the Del Carmen Formation, 165 feet thick.

Breccia pipes develop when limestone dissolves in water creating caverns; the caverns collapse and fill up with breccia, a mixture of broken rock and rubble.

Mining in the area was first reported in 1894. By 1898 a settlement had grown up around the mining activity and a post office called Terlingua had opened. By spring 1900, two other mines had begun operations and a mining boom was under way.

Howard Perry, a Chicago businessman, entered the business after gaining title to land at Long Draw Graben through non-payment of a debt. When he received more than one offer for the property, he sent an agent to investigate, and found that it bordered some productive cinnabar properties. He formed the Chisos Mining Company in 1903 and it very quickly became the dominant mine, thanks to the biggest and most valuable ore body in the area, a 75 foot diameter breccia pipe found underground at a depth of 550 feet and mined from there down to 840 feet.

## Fossil Knobs

**Left** Much of the richest cinnabar ore was found along the Fossil Knobs ridge above Long Draw. Waste tips and abandoned mining machinery now litter the ridge; those on the right are from the Chisos Mine, those on the left from the Rainbow Mine. The latter began operating in 1916, during an upsurge in demand fueled by the First World War. After a long struggle, rife with chicanery, it was taken over by the Chisos Mine in 1938.

The detailed history of Terlingua mining is told in *Quicksilver: Terlingua and the Chisos Mining Company* by Kenneth Ragsdale, published by Texas A&M Press.

Photographed from Mile 5.8.

Mercury prices slumped when the First World War ended in 1918 and demand for mercury fulminate plummeted. The Chisos Mine experienced a slow decline through the 1920s and the Depression years of the 1930s, never regaining its wartime prosperity. The mine finally ran out of ore in 1942 and the company filed for bankruptcy. Although a company owned by the Brown brothers of Houston, founders of Brown and Root, bought it and spent heavily on exploration, they failed to find more ore, and the mine never reopened. The Fresno Mine, seven miles west of Terlingua, and the 248 Mine at Mile 2.1, both opened in 1940 when the onset of the Second World War created a flurry of activity. The former continued in business until 1972 and was the last mine to close.

Terlingua was a ghost town until tourism brought new life to the area in the late 1960s. It became famous for its annual chili cook-off and in 1967 was christened the "Chili Capital of the World" by the Chili Appreciation Society. The old company store reopened as a gift and art shop, river float trips are scheduled in the former cantina and a dinner theater occupies the former motion picture theater. The combined Study Butte-Terlingua population in the 2000 census was 267.

At Mile 4.9, a half-mile beyond the ghost town exit, the road drops steeply into Long Draw Graben, descending 125 feet down the fault scarp of the graben's northeast boundary. The graben is 6 miles long and a half-mile wide with strata on this fault dropping between 500 and 1,000 feet down into the graben. Fossil Knobs (3,234 feet), the ridge above the escarpment at 3 o'clock, contains a buried igneous intrusion that has baked and hardened the surrounding limestone, making it resistant to erosion.

The road traverses the Long Draw graben diagonally for about 2 miles, crossing the draw three times. On the left, the southwest boundary fault of graben runs along the base of the mesa called Reed Plateau with a displacement of 1,600 feet down towards the road. Light colored Buda and Del Rio sedimentary rocks crop out on the escarpment with steeply dipping thick dark beds of Santa Elena limestone below. Boquillas flaggy limestones crop out in the graben floor on the right.

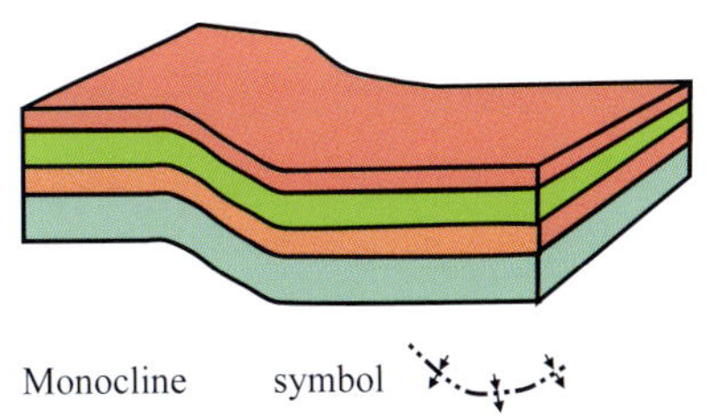

Monocline symbol

### The Terlingua Monocline

At Mile 7.4, the road climbs 270 feet up the southwestern boundary fault scarp on to Thirty-eight Hill. The fault here has a displacement of 2,000 feet. At Mile 7.7, the road descends 180 feet back into the Castolon Graben through Santa Elena limestone road cuts. The limestone cliffs on both sides of the road dip steeply down the Terlingua Monocline. A monocline is a term geologists use for a flexure or local steepening of strata, as illustrated by the drawing on the left.

### Flatirons on the Terlingua Monocline

**Above** Looking back from Mile 11.4, thick flatirons of Buda limestone dip down the Terlingua Monocline at the Sierra de Cal.

A flatiron is a short, triangular hogback forming a ridge or spur on the flank of a hill that looks like a flatiron. A flatiron is usually a plate of steeply inclined resistant rock, in this case limestone.

The vertical drop at Sierra de Cal on the surface is about 400 feet with the monocline continuing several hundred more feet underground. Geologists believe that the monocline is the result of strata being draped over an ancient fault.

Once down the monocline, the road continues through rolling Boquillas limestone hills to Lajitas. Rainfall is sparse in this area, averaging 8.9 inches per year, only enough for a few shrubs to grow on the sandy, pebbly soil and the limestone residue.

### Big Bend Museum of Paleontology and Geology

At Mile 10.7, the Big Bend Museum of Paleontology and Geology exhibit is on the right. The exhibit houses hadrosaur, mosasaur, ceratopsian and Ptychodus fossils collected by Ken Barnes from Upper Cretaceous strata in the Big Bend area. The exhibit is open by appointment.

Hadrosaurs (meaning "bulky lizards") were a family of duck-billed, herbivorous dinosaurs, the most common dinosaurs, ranging in size from 10 to 40 feet long and weighed up to five tons.

## Sierra Aguja

**Above** Erosion in Castolon Graben is controlled by Terlingua Creek and has now reached a level where the Chisos strata has been mostly removed and Aguja, Pen and Boquillas sedimentary rocks are under attack. As a result, volcanic strata are absent except for Sierra Aguja (above) (*Needle Peak* in Spanish) where a Tule Mountain Trachyandesite cap has protected the underlying 700 feet of Chisos tuffs and lavas. Chisos Formation light gray tuffs underlie the lava, some of which can be seen outcropping on the peak's flanks. This is a similar sequence to North Lajitas Mesa, so it seems likely that Chisos strata were originally continuous across the graben. The Tule Mountain unit may not have spread over the entire Castolon Basin.

Photographed from Mile 9.6.

The group probably evolved in central Asia and by the Late Cretaceous (98 to 65 Ma) had spread all over the lands of the northern hemisphere, migrating across a land bridge that existed at that time connecting Asia to North America. From there, the group moved eastward again to Europe. One type of hadrosaur fossil has even been found in southern Argentina, having migrated from North America across across a volcanic island chain that once existed.

The outward appearance of hadrosaurs varied widely amongst the different species but all had similar skeletons. The most obvious common feature was the elongated face ending in a broad, flattened snout. The horny, toothless beak of the hadrosaur looked quite similar to the bill of a duck, hence the common nickname "duck-billed dinosaur" for these dinosaurs,

Hadrosaurs had hoof-like nails on their feet, and bumpy skin. They ran on two legs, balancing themselves with their heads and stiff tails and may have walked on all four legs while grazing. Hadrosaurs probably lived near bodies of water, migrating to high ground to lay eggs.

## Mosasaur

**Left** The Onion Creek, Travis County, mosasaur (*Mosasaur maximus)* was found in 1934 by geology students Clyde Ikins (pictured) and John Smith of the University of Texas. The Onion Creek Mosasaur belonged to one of the larger species of mosasaurs, and one that lived only a short time before the last mosasaurs went extinct. With whale-like flippers instead of feet, and a large, flattened tail, these giant creatures probably swam with a snake-like motion, using the tail as a propeller and rudder, and the flippers as stabilizers. Their long slender bodies probably made for great speed and agility. The Onion Creek Mosasaur is 30 feet long, about 12 feet of which are tail. The head is 4 feet 8 inches long, and the jaws, when fully opened, have a gape of about 3 feet.

Photograph and description courtesy of the Texas Memorial Museum, Austin.

## Tricerops

**Left** Artist's reconstruction of *Tricerops,* a ceratopsian dinosaur.

## Tricerops

**Left** The side view of a Tricerops skeleton, Senckenberg Museum, Frankfurt, Germany.

**Hadrosaur**

Artist's rendering of a hadrosaur. Note the flattened snout and the heavy tail. The tail was probably used to balance the animal when it ran.

Image: National Geographic Channel.

Hadrosaurs were herbivores (plant-eaters) and only had teeth in the cheek area, not the front of the mouth. The many rows of teeth were perfect for effectively grinding up rough vegetation for easy digestion.

Mosasaurs were not dinosaurs but lepidosaurs, reptiles with overlapping scales. These predators evolved in the Early Cretaceous Period and during the last 20 million years of the Cretaceous Period (85 to 65 Ma) became the dominant marine predators. Mosasaurs breathed air and were powerful swimmers that were well-adapted to living in the warm, shallow seas that overlay continents in the Late Cretaceous. Mosasaurs gave birth to live young, rather than return to the shore to lay eggs, as sea turtles do.

Ceratopsia is a group of herbivorous, beaked dinosaurs which thrived in what are now North America and Asia during the Cretaceous Period, although ancestral forms lived earlier, in the Jurassic (artist's rendition above). Early members were small and bipedal. Later members, including ceratopsids like Centrosaurus and Triceratops, became very large quadrupeds and developed elaborate facial horns and a neck frill. While the frill might have served to protect the vulnerable neck from predators, it may also have been used for display, thermoregulation, the attachment of large neck and chewing muscles or some combination of the above. Ceratopsians ranged in size from 3 feet and 50 lbs. to over 30 feet and 12,000 lbs.

Ptychodus is a family of sharks with teeth adapted for crushing shellfish. As their skeletons were made of cartilage rather than bone, only their teeth fossilized well and are found widely around the world, especially in North America. Sharks don't have a single row of teeth. New ones constantly replace or push older ones to the side making the teeth row very wide.

No one is quite sure what Ptychodus looked like, but shark-like vertebrae suggest that they may have been more shark-like than ray-like. Judging from the size of some isolated teeth, these fish probably grew to lengths of 12-15 feet. Ptychodus became extinct in the Upper Cretaceous, about 86 Ma.

## North Lajitas Mesa

North of FR 170, North Lajitas Mesa (3,672 feet) rises 1,300 feet above the highway. It and its twin, South Lajitas Mesa (3,069 feet; photograph on page 80), are blocks of Chisos Formation lavas and tuffaceous sedimentary rocks resting on Pen Formation claystones. This view of the north mesa photographed from Mile 16.2, shows the Tule Mountain Trachyandesite lava cap with its surrounding apron of pinkish broken boulders, overlying pink and white tuffs. A thin bed of Bee Mountain Basalt crops out

on the left of the mesa. A thick bed of black Alamo Creek Basalt crops out on the right. Near the base of the mountain you can see the red beds of the lowest Chisos Formation, which we will see again in Fresno Creek valley. Below them are yellowish gray calcareous marine clays of the Upper Cretaceous Pen Formation at the base of the slope, and below them, flaggy Boquillas Formation limestones in the foreground. Aguja Formation strata are missing here.

### Lajitas

The entrance to the Lajitas Resort (2,364 feet) is at Mile 16.6. The setting, on a bluff overlooking the Rio Grande at the San Carlos ford of the old Comanche Trail, is arresting. South of FR 170, the emerald green golf course on the river flood plain is in striking contrast to the exposed Boquillas mesas and mounds above it. Behind the mesas, the northern tip of Mesa de Anguila is folded steeply and dips 850 feet into the flood plain on the west. Basalt sills in the Santa Elena limestones parallel to the bedding emphasize the fold.

The name Lajitas is Spanish for *little flat rocks* from the Boquillas flagstones that crop out in the vicinity. Lajitas was made a port of entry into the United States in 1900 when border traffic increased because of mining activity in Terlingua and Mexico. Farming along the narrow flood plain of the river brought in more families, and by 1912, the town had a store, a saloon, a school with fifty pupils, and a customhouse. The crossing, a smooth rock bottom across the river, was the best between Del Rio and El Paso.

By 1949, with the closing of the Terlingua mines, the population had dwindled to four and so it remained until Walter Mischer of Houston bought the area in 1977. He developed and restored the settlement and when he sold it in 2000, Lajitas was a resort with seventy-five residents and fifteen businesses. The old church had been restored and the trading post remained open. There was a hotel, a restaurant, a 9-hole golf course, a swimming pool, an RV park, and an airstrip. Between 2000 and 2007 when the resort changed hands again, the golf course was extended to 18 holes, a restaurant, an equestrian center and various other amenities were added, and a substantial acreage was opened up for residential development.

**South Lajitas Mesa**

One of the best views of South Lajitas Mesa (3,069 feet) is from the edge of the bluff. The mesa is a block of Chisos Formation lava and tuffaceous sedimentary rocks resting on Pen Formation claystone and capped by Tule Mountain Trachyandesite.

Photographed from Mile 16.7.

## Big Bend Ranch State Park

FR 170 enters Big Bend Ranch State Park at Lajitas. The largest state park in Texas was opened to the public in 1991. It comprises 280,280 acres north and west of Lajitas. The Barton Warnock Education Center at Mile 15.4 just before Lajitas has an excellent botanical garden and bookstore. Another visitor center is located at Fort Leaton at Mile 62.5.

Among the park's geological attractions are volcanic craters in the Bofecillos Mountains and the Solitario. Facilities include picnic tables at Madera Canyon, rafting and canoeing along Colorado Canyon, several trails, and an unimproved primitive campground. The area has been declared an international biosphere reserve by the federal government. It is home to eleven endangered species of plants and animals and ninety major archeological sites. Almost 400 species of birds either live in or migrate through the area.

## The Rio Grande Rift Segment along the River

The Rio Grande rift segment continues along the river from Lajitas, narrowing as it reaches the Bofecillos Mountains. Very little extension has taken place in the narrow section of the river valley between the Bofecillos and Sierra Rica volcanic centers; extension is usually limited around volcanic centers. Geologists think that the underlying solidified magma chambers act as buttresses that resist movement.

The corridor between the two volcanic centers is heavily faulted. It is what geologists call a *transfer fault zone.* The Tascotal Mesa Fault, which runs across the Bofecillos block farther north, is another such zone. Transfer fault zones are one of the ways in which fault blocks adjust to the crust being stretched. In this case, as the Rio Grande rift segment developed and the crust was stretched from southwest to northeast, the block south of the Tascotal Mesa Fault, rather than move to the northeast, moved about a half-mile to the

**Left** Looking up the Rio Grande flood plain to blocks of Chisos strata in Mexico dipping quite steeply towards the river. The gravel road leads to the San Carlos ford of the old Comanche Trail, a crossing which is no longer in use. The volcanic hills of Sierra el Mulato are on the skyline at right. A few buildings in the Mexican village of Paso Lajitas can be seen on the far left.

Photographed from Mile 17.2.

## South Lajitas Mesa

Looking up at South Lajitas Mesa from the highway on the way down to Comanche Creek. The basal cliff is of Chisos tuff and the cap is Tule Mountain Trachyandesite. Midway up the slope, ledges of yellow-brown Mule Ear Spring Tuff and dark brown Bee Mountain Basalt crop out, with Chisos Tuff above and at roadside. Comanche Creek is in the foreground.

Strata are about 300 feet lower here than in North Lajitas Mesa because of a fault running up the valley of Comanche Creek.

The tuff beds dip at a low angle towards the river. From this point, many of the blocks on both sides of the river are tilted.

Photographed from Mile 16.9.

west compared to the block to its north. We know this because an intrusion just north of the Solitario was cut in two by the fault and its southern half is now a half-mile west of its northern half. We do not know how much movement took place along the river transfer fault zone.

Thus Rio Grande rifting in this area is characterized by extension in places, such as in the Sunken Block and the Presidio Graben underlying the Presidio Basin. In other places, extension is combined with transverse or strike-slip faulting as we see along this corridor and in the Redford and Santana Grabens.

The amount of extension in the Presidio Graben has been calculated at between one and two miles. The distance is calculated from the angles of the boundary faults and the distance strata have moved down along them. Extension in the Redford Graben is somewhere between a half-mile and a one mile.

The Earth's crust is broken into blocks along the rift segment, many of which are tilted, some spectacularly so, as you can see in the photographs on pages 82, 84, 95, 97, 99, 114, and 117. In all these photographs, blocks can be seen tilted towards or away from the river valley.

The river normally follows the lowest blocks but not always. In some places it follows a path formed at a higher level; it was *superimposed.* A good example is Colorado Canyon, where the river cut through Colorado Mesa, a block of Santana Tuff more than 1,000 feet thick. There, the river had set its course when the ground level was much higher than it is today. As erosion removed graben fill and carried it downstream, the river continued its course rather than being diverted to the low block north of the mesa (photograph on page 102). Closed Canyon which cuts through Colorado Canyon is another example of a superimposed stream (photograph on page 104).

As well as being tilted, blocks of strata along the river are consistently some 2,000 feet lower than they are north of the river. The blocks step down into the rift segment through west-northwest faults, most of which are plotted on the geological map, continuing the pattern we saw coming down TX 118 into Study Butte. An example of such stepped-down strata is the Mitchell Mesa Rhyolite capping the flat-topped bench (2,720 feet) running east from Colorado Mesa (photograph on page 102). The formation here is almost 2,000 feet below its level along Green Valley, on Bandera Mesa (4,675 feet), for example. Other Mitchell Mesa Rhyolite outcrops can be seen near river level as far as the eastern boundary of the Presidio Basin at Mile 56.8 (photograph on page 117). Stepped-down strata continue up the river valley at least as far as El Paso.

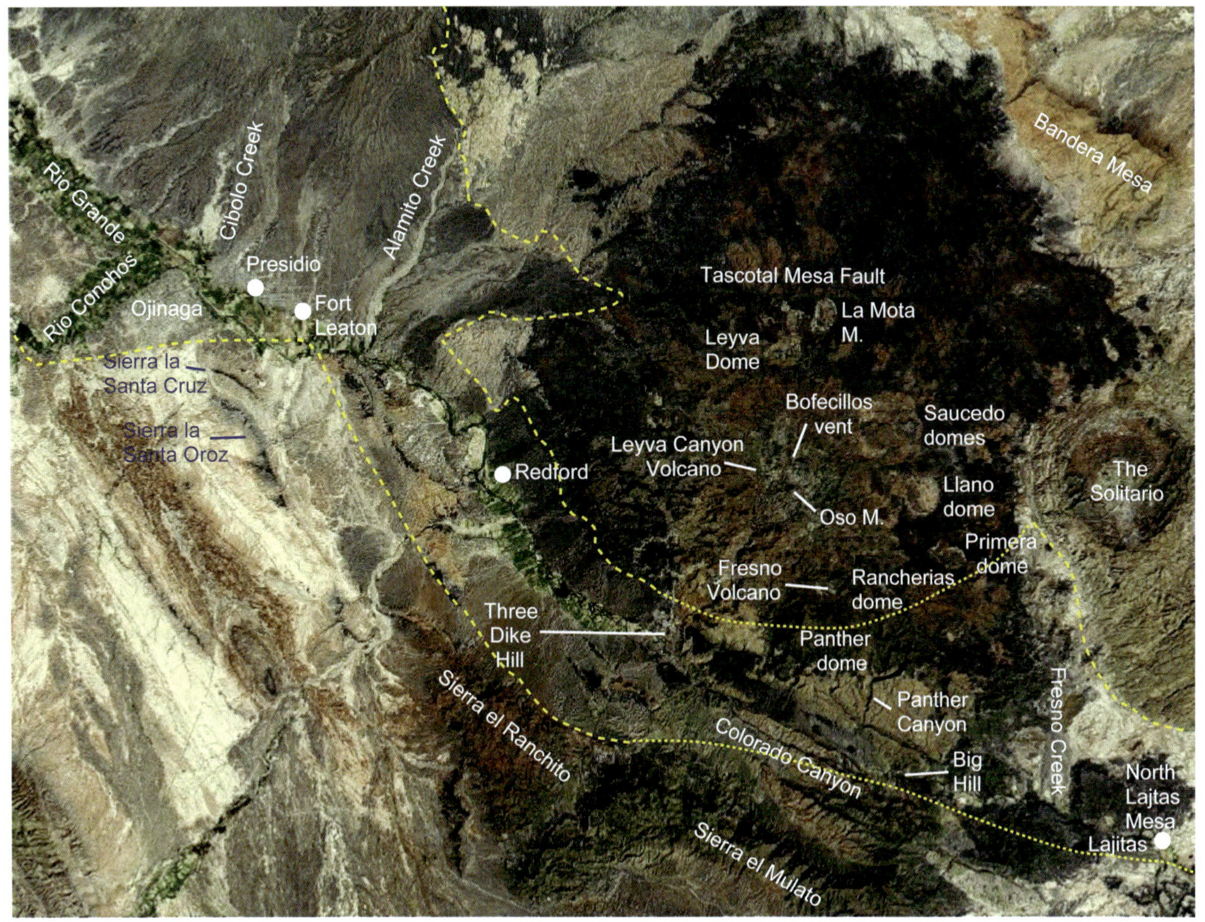


**The Satellite View**

The main features of the area can be seen in the satellite photograph above. The Rio Grande rift segment boundaries are outlined with yellow lines, dashed where known, dotted where less certain.

Beginning at Lajitas, North Lajitas Mesa and the hills to its left as far as Big Hill are capped by Chisos Formation strata, mainly Tule Mountain Trachyandesite, interrupted only by the broad valley of Fresno Creek west of the Terlingua Monocline. Just to the left of the label for Fresno Creek on the satellite photograph, the yellow-brown body is a large intrusion, with the circular light-colored Contrabando Dome to its southeast where Boquillas limestones have been domed by an underlying intrusion. You can see the dome and the Solitario looking north from the Fresno Creek bridge at Mile 23.0.

At Big Hill at Mile 29.8 (not to be confused with the Big Hill just south of Alpine), the Chisos Formation lavas and tuffs disappear

**Left** Chisos Formation strata along the east side of Fresno Flat, photographed from Mile 22.4. The sequence of strata on display is similar to that on South Lajitas Mesa – a thick bed of Tule Mountain Trachyandesite as caprock with beds of yellow brown Mule Ear Spring Tuff, dark brown Bee Mountain Basalt, and white Chisos tuffs, red near the base of the formation. These red tuff beds are also seen at roadside at Mile 21.3 and in North Lajitas Mesa (page 78).

under San Carlos and Santana welded tuffs from the Santana and San Carlos Calderas south of the river. The San Carlos Tuff, slightly darker than the Santana Tuff, appears in Dark Canyon, crosses the road and crops out on Santana Mesa as far as the La Cuesta River Access at Mile 30.7 (photographs on pages 105 and 109). Its boundary is probably that of the San Carlos Caldera.

Several Bofecillo volcanoes can be seen in the satellite photograph. The Rancherias Dome is the site of the Fresno Volcano, the oldest volcano in the field, which erupted trachyte and rhyolite lavas 32 Ma, with lavas flowing away from the vent like spokes of a wheel. A substantial cone built up around the vent, now only 200 feet high as a result of erosion. The Rancherias Trail, at Mile 34.8, cuts through the cone, a 19-mile round trip.

The Fresno Volcano eruption was followed by six others that erupted basalt and rhyolite lavas and ash-flow tuff from about 27.3 to 27.1 Ma. The rhyolite lavas created domes, such as in the Leyva Canyon Volcano, which built up a large cone of rhyolite lavas, tuffs and sedimentary rocks around its vent, the Bofecillos vent, which is still the highest and most rugged part of the Bofecillos Mountains. An aureole of light pink lavas and tuffs around the volcano contrasts with the dark basalt lavas surrounding it.

The inner section of the aureole is a circular basin of gray-green material half a mile wide around the vent. The basin has a rim of syenite intrusive rocks and is filled with a coarse breccia of igneous rocks from the rim of the vent, some of them as large as 500 feet in diameter. Oso Mountain (5,135 feet), a syenite intrusion on the rim is the highest point in the Bofecillos Mountains.

Basalt lavas created broad plateaus such as the Sauceda Volcano, which soon after the Leyva Canyon eruption began erupting basalt and trachyte lavas. These lavas spread through the mountains in a series of thin flows that you see as the dark areas in the photograph, from north of the Solitario to north of Redford and south as far as Three Dike Hill, which is capped by one of the basalt lavas. The lavas came from widespread fissures, three of which are seen on Three Dike Hill.

Another widespread volcanic rock is the dark red-brown Rancho Viejo Tuff which erupted in multiple pulses and is found in several layers, 3 to 30 feet thick, throughout the mountains from Fresno Canyon to Three Dike Hill. Its source is uncertain but is most likely the Bofecillos vent.

A number of domes can be seen on the satellite photograph. Most are remnants of younger rocks or *outliers* such as the Panther, Segundo, Llano, Saucedo domes. Others, such as the Primera and Leyva Domes and La Mota Mountain, are intrusions.

Terneros Creek runs along the Tascotal Mesa Fault across the mountains just north of La Mota Mountain, following a valley created by the creek along a zone of rocks fractured by the fault.

The Tascotal Mesa Fault continues to the west just south of Redford where it forms the southern boundary of the Presidio Graben and Basin. The basin is silvery gray in Texas but more creamy yellow in Mexico. South of Redford the dark gray Redford Basin straddles the Rio Grande, the latter easily identified by its border of green irrigated fields.

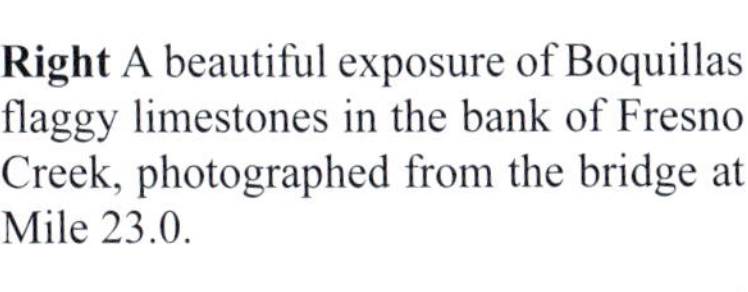
**Right** A beautiful exposure of Boquillas flaggy limestones in the bank of Fresno Creek, photographed from the bridge at Mile 23.0.

The photographs on these two pages and the next three pages are taken from the broad flat of Fresno Creek, which runs up from the Rio Grande for about five miles, mostly on Upper Cretaceous Pen and Boquillas Formation strata. Beyond that point, Fresno Creek has cut Fresno Canyon into volcanic strata.

## The Solitario

The Solitario is a 10-mile diameter dome with a resistant rim of Lower Cretaceous limestone dipping outward radially. Doming was caused by a 4-mile diameter laccolith being intruded at an original depth of 2½ miles. Cretaceous strata are older as you go towards the center of the dome. Fresno Peak (5,120 feet) in the upper left is in Del

Carmen Formation limestone, 925 feet thick on the east of the dome. The flatirons on the outside are of the younger Santa Elena Limestone, which is about 800 feet thick on the east of the Solitario. A flatiron is a short, triangular hogback forming a ridge or spur on the flank of a hill that looks like a flatiron. A flatiron is usually a plate of steeply inclined resistant rock, here light gray Santa Elena Limestone.

Photographed from just beyond Fresno Creek bridge at Mile 23.0.

The beautiful cliffs along the west side of the Fresno Creek valley from Mile 22.1. The uppermost reddish bed is Tule Mountain Trachyandesite, followed by Chisos Formation

tuffs, dark Bee Mountain Basalt, more tuffs, and on the flat in front, yellowish gray Cretaceous Pen Formation marine clay.

### Fresno Creek Valley from the West

This photograph, taken from Mile 24.1 at nightfall, looks back at the Fresno Creek valley through banks of gravelly alluvium. The valley is a flat-bottomed erosional basin stretching several miles north into Chisos Formation volcanic strata. The succession on the cliffs opposite is plainly visible – a thick bed of Tule Mountain Trachyandesite is underlain by Chisos tuffs with several thin beds of Bee Mountain Basalt and Alamo Creek Basalt interbedded.

The trachyandesite pillar on the left skyline can be seen for several miles to the west. It stands above the "Contrabando" movie site at Mile 21.5. The site has several adobe buildings and a stone wall. It was last used for the 1995 made-for-TV movies "Streets of Laredo" and "Dead Man's Walk." Although a movie called "Contrabando" exists, it was made in Mexico; the set takes its name from Contrabando Creek which crosses the highway at Mile 21.6.

Notice how the block on the skyline at mid-picture tilts towards the river. Mesa de Anguila is on the skyline to its right. At far right, another block also tilts in the same direction, in this case away from the river.

## Chisos Formation Tuffs

Chisos tuffs are prominent along the river from the west of Fresno Creek valley to the Tepees picnic area at Mile 28.7.

**Above Left** A pyramid at Mile 24.5 is capped by a remnant of Tule Mountain Trachyandesite with a thick Bee Mountain Basalt bed in mid-face. A small fault to the left of the peak has dropped strata down to the left.

**Below Left** A roadside tuff bed at Mile 27.4 has been eroded into bizarre shapes by wind and water. Such pillars are called *hoodoos* in the southwest.

**Below** Lava beds dipping steeply away from the river cap a fine exposure of Chisos tuffs on the Mexican side of the river. The cliffs of Sierra el Mulato are on the horizon. Photographed from Mile 27.1 at a wide spot in the valley.

Looking east from Big Hill in the evening light, the dark chocolate-brown boulders in the foreground are of San Carlos Tuff. The source of the tuff, the San Carlos Caldera, crosses over the river here for a short distance. This photograph gives a good portrayal of the river winding along between volcanic blocks.

The Chisos Mountains are on the left center skyline with Emory Peak the high point. To the right of the Chisos Mountains, the pillar on the skyline is a prominent landmark just above the Contrabando Movie Set.

Photographed from Mile 29.9.

## Dark Canyon

**Above** The Rio Grande has cut the 700-foot deep Dark Canyon through dark San Carlos Tuff at Big Hill.

Flows of San Carlos and Santana Tuffs probably dammed the river valley here when they first erupted. The dam would have been at least as high as the highest point on Santana Mesa ahead, 4,000 feet above sea level, and 1,680 feet above the river level today.

Photographed in the evening light from Mile 29.2.

### Strata from Big Hill to Three Dike Hill

| Period/Epoch | Formation | Age | Description |
|---|---|---|---|
| Oligocene | Rawls | | Two members of this formation crop out along the River Road: |
| | | 27.1 | **Sauceda Lava Member:** Quartz trachyte, trachyte and basalt lavas, porphyritic in part, erupted from the Sauceda Volcano. |
| | | 27.0-27.3 | **Leyva Canyon Member:** Tuff, densely to poorly welded, coarse fluvial conglomerate and coarse debris-flow deposits, overlying quartz trachyte and rhyolite lavas and rhyolitic ash-flow tuff at base; erupted from the Leyva Canyon Volcano. |
| | Santana Tuff | 27.8 | Densely-welded to non-welded rhyolite ash-flow tuff; erupted from Santana Caldera; up to 600 ft. thick along the Rio Grande; thinning and pinching out into the Bofecillos Mountains. |
| | San Carlos Tuff | 30.5 | Densely-welded rhyolite ash-flow tuff; more than 650 ft. thick at Big Hill; pinching out within 0.6 mile north against what is probably the north edge of its source, San Carlos Caldera; may all be caldera fill. |
| | Fresno | 32.0 | Trachyte and rhyolite lavas at base followed by an interval of conglomerate and sedimentary volcaniclastic rocks, with basalt lava at top. |

### San Carlos and Santana Tuffs

The highway continues among cliffs and outcrops of Chisos Formation strata until just before the Tepees picnic area at Mile 28.7, where cliffs of Santana Tuff can be seen north of the road overlying the Chisos strata.

The Santana Tuff, one of the major volcanic units seen from here to Tapado Canyon at Mile 41.1, originated in the Santana Caldera in the Sierra Rica about 27.8 Ma. The north rim of the caldera is about 10 miles south of FR 170.

The San Carlos Tuff outcrops under the Santana Tuff on Big Hill and along both sides of the Santana Basin ahead. It erupted from the San Carlos Caldera at 30.5 Ma, three million years before the Santana unit and is largely buried by it. In outcrop it is a rather darker brown than the Santana Tuff.

Both tuffs are strongly welded along the river near their sources. Welding occurs when very hot volcanic ash lands and re-melts welding the ash particles together into a welded tuff. Sometimes the welding is so complete that the rock is very difficult to distinguish from lava.

The top of the San Carlos Tuff, at about 2,840 feet above sea level, can be seen on both sides of the river from the overlook on Big Hill. The tuff crops out about 1,100 feet higher on Santana Mesa ahead because of a fault running along the edge of the mesa that drops strata down into the Santana Basin.

## Santana Basin

**Above** Coming over the top of Big Hill at Mile 29.8, the expanse of the Santana Basin, seven miles long and a mile and a half wide, is surprising after the cramped river valley. The basin is underlain by a graben of unknown depth filled with sandstone, clay and conglomerate, Petan Basalt and alluvial fan deposits. The southern boundary fault of the graben underlying the basin, marked in white on the diagram, runs along the right of Colorado Mesa and across to the outcrop of white Chisos tuff.

The Rio Grande, rather than follow the low ground to the right of Colorado Mesa, has carved the 1,000 foot deep Colorado Canyon between the mesa and the cliffs to its left, an example of a superimposed stream (see page 83). A narrow mesa capped by Mitchell Mesa Rhyolite continues the canyon towards the camera.

Faulting and erosion have provided a cross section of a rhyolite lava dome in the cliffs on the left. The columnar jointed dome shows how viscous rhyolite magmas form thick flows that do not travel far from their sources. San Carlos and Santana Tuffs filled in around the dome. At center, an ancient rock fall has created a white trail, a landslide scar, down the cliff.

Photographed from the western slope of Big Hill at Mile 29.9.

## Panther Canyon

**Below** Panther Canyon cuts through Santana Tuff on Santana Mesa. The canyon is over 700 feet deep at the exit from the mesa. The creek bed meanders across the basin about a mile in front of Big Hill and crosses FR 170 at Mile 31.5.

The hills seen through the gap are of the Panther Dome, in which Chisos tuffs, Tule Mountain Trachyandesite and Mitchell Mesa Rhyolite have been domed up, presumably by a hidden intrusion. The Mitchell Mesa unit here is about 3,800 feet, midway between its elevation on Bandera Mesa and in Santana Basin.

Photographed from Mile 34.0 in the evening sun.

## Closed Canyon

**Above** In Closed Canyon, what is a relatively minor stream today has cut a 350 foot deep chasm through the Santana Tuff of Colorado Mesa. This is an example of a superimposed stream, one that had set on its course at a higher level when the basin was full of debris from the surrounding hills. Once set on its course, it continued cutting down through the mesa.

Photographed at Mile 36.4.

## Santana Mesa

**Above Right** San Carlos and Santana Tuffs on Santana Mesa from La Cuesta River Access at Mile 30.7. The rhyolite lava below the tuffs is in the Fresno Formation, and is probably from the same source as the rhyolite dome across the river.

## Santana Tuff

**Below Right** This sheer jointed Santana Tuff cliff is on the right at Mile 38.7 just before Tapado Canyon.

**Following Page** This beautiful Santana Tuff cliff was photographed just beyond the Colorado Canyon Access at Mile 38.0. Nearly continuous cliffs of this welded tuff line the highway for the next three miles.

San Carlos Tuff
Santana Tuff
Rhyolite

## Spring Flowers

The spring flowers come out when it rains.

**Above left** The Big Bend Bluebonnet (*Lupinus havardii* S. Wats.) is more spindly than the common Texas bluebonnet.

**Below Left** Prickly pear (*Nopal opuntia*).

**Above** Trumpetbush (*Tecoma stans* (L.) Juss. ex Kunth) on the east side of Big Hill. The dark San Carlos Tuff crops out in the foreground.

**Right** Hedgehog cactus (*Echinocereus enneacanthus*).

**Looking up Tapado Canyon**

The high bluff on the right is capped by basalt and trachyte lava from the Sauceda Volcano, followed by debris flow deposits of the Leyva Canyon Volcano. The ledge is capped by Santana Tuff followed by Fresno Formation trachyte and conglomerate, tuff sandstone and pumice tuff from the Fresno Volcano.

This is the last exposure of Santana Tuff along the River Road. The sequence can also be seen in the photograph on the following page; the bluff is on the skyline above the highway.

Photographed from Mile 41.2.

## The Eastern Margin of the Redford Basin

The magnificent panorama above on the eastern boundary of the Redford Basin begins with Three Dike Hill (3,430 feet) on the left. The mountain has a dark cap rock of basalt lava from the Sauceda Volcano. Underneath is a very distinctive light-colored bed of coarse sedimentary deposits of eroded ash-flow tuff and other volcanic rocks from the Leyva Canyon Volcano followed by a thin brown bed of basalt lava, also from the Leyva Canyon Volcano. Under the basalt is a series of light gray sedimentary rocks – Fresno Formation conglomerate, tuff sandstone and pumice tuff from the Fresno Volcano, well displayed to the right of the third dike.

Three Dike Hill gets its name from the three dark dikes that run down the center of the face. The dikes are the solidified remains of magma that ascended 30 to 60 miles through fractures to feed the upper Saucedo lava flow. The dike on the right continues down

diagonally through the sedimentary rocks and across the photograph. The two other dikes are covered up by debris below the Leyva Canyon basalt bed.

The block in the right foreground, with the same strata sequence as Three Dike Hill, has been down-faulted some 700 feet and tilted to the left, another example of tilted and stepped-down strata along the Rio Grande Rift. Immediately to the right of the road in front but not in the Redford Graben, Saucedo basalt lava crops out, 1,900 feet below outcrops on the hills to the north, showing that the strata along the river continue to be down-faulted by about this amount.

The high plateau (4,121 feet) on the skyline to the right of Three Dike Hill is capped by Rawls Formation lavas and tuffs somewhat younger than those on Three Dike Hill.

Photographed looking east from a crest in the highway at Mile 43.5. (See also close up on page 6 of Three Dike Hill from the roadside at Mile 42.5).

## The Redford Basin

This photograph from Mile 47.1 takes in the full panorama of Sierra el Mulato and Sierra Rica across the Mexican section of the Redford Basin. The Rio Grande is directly in front of the camera and can be seen on the left foreground of the photograph.

At left center, a tilted block of volcanic beds is capped by Mitchell Mesa Rhyolite. Behind the tilted beds is the rugged profile of Sierra el Mulato (about 5,500 feet), composed of lavas younger than the Mitchell Mesa Rhyolite.

Near the river on the left, a thick bank of horizontal alluvium can be seen in front of the tilted block. Many examples of alluvium lying at an angle to older strata can be seen in the Redford and Presidio Basins.

Sierra Rica (7,500 feet) is on the skyline on the right. This mountainous range, at one time mined for silver (hence the name "Rich Mountain") is in the Santana caldera and is built up of volcanic rocks that erupted from that caldera. The photograph on page 128 shows the mountain from near Presidio.

The eastern flank of the ridge Sierra el Ranchito ridge (4,400 feet), begins behind the houses on the right.

### The Redford Basin

Just beyond Three Dike Hill at Mile 42.5, the highway enters the Redford Basin and the river valley opens up dramatically. Two thirds of the basin, which is 15 miles long and six miles wide, are on the Mexican side of the river. Occasional outcrops of Sauceda basalt lava are found in the first mile after Three Dike Hill. The unit occurs at roughly the same level as in front of Three Dike Hill, which is 1,900 feet below its position farther north. Beyond that point, the graben deepens, although by how much is unknown as no oil wells have been drilled in it.

Views across the Redford Basin are some of the most spectacular along the River Road as the photographs on pages 112-123 attest.

**Right** Steeply dipping Chisos Formation tuff and sandstone capped by Mitchell Mesa Rhyolite crop out on the left of the road at Mile 56.8. These beds mark the eastern boundary of the Presidio Basin.

The Mitchell Mesa unit here is about 2,700 feet above sea level, 2,000 feet below its elevation on Bandera Mesa (4,675 feet).

### Redford Basin

**Left** The Rio Grande winds its sluggish way through the Redford Basin in this photograph taken from a bluff overlooking the river at Mile 47.1.

Several small hamlets border the irrigated crop lands in Mexico between this point and Redford.

## Redford

The small hamlet of Redford straddles the road at Mile 49.8. The name comes from the red rock at an old ford on the Rio Grande. The population of the settlement was given in the 2000 Census as 132. The main occupation is farming on irrigated land along the river flood plain.

At about Mile 55.0, the highway leaves the Redford Basin and crosses a three mile wide section of tilted lava and alluvial banks as shown in the photograph below.

## Cerro de las Burras

The long ridge of Cerro de las Burras (4,345 feet) is on the northern boundary of the Redford Basin. The sequence of beds is much the same as on Three Dike Hill except for three younger units at top – a Petan Basalt lava bed, and a trachyte lava and a trachytic ash-flow tuff, both 27.1 m.y. old.

Below these beds are basalt lavas from the Sauceda Volcano, light-colored coarse sedimentary deposits of eroded ash-flow tuff and other volcanic rocks from the Leyva Canyon Volcano, followed by a thin brown bed of basalt lava, also from the Leyva Canyon Volcano, and at base a series of light gray sedimentary rocks – Fresno Formation conglomerate, tuff sandstone and pumice tuff from the Fresno Volcano.

The long mesa to its right, tilted to the north, is on the eastern boundary of the basin. It is made up of a similar sequence of rocks, beginning with Rawls Formation basalt lava. Light yellow gray Fresno sedimentary rocks crop out at its base.

Strata in Cerro de las Burras, farther back from the river valley, are not tilted

Photographed from Mile 48.9.

### The Redford Basin

**Above** In the Redford Basin, the road runs between gravel banks of graben fill to the north, as seen in the photograph above, and the river valley to the south. The desert willow in the foreground blooms with delicate pinkish flower in the springtime.

**Left** Like desert everywhere, flowers bloom in abundance when the rains arrive. The photograph opposite shows pink prickly pear in front with the more common green prickly pear in flower at back.

### An Attractive Vista over the Redford Basin

In this panoramic view looking back over the Redford Basin, the ridge on the left has Fresno tuff overlying Mitchell Mesa Rhyolite and Chisos Formation white tuff. Cretaceous Boquillas limestone creates a yellow cliff in mid-photograph, the only exposure of this formation since Fresno Creek at Mile 23.0. The multicolored volcanic rocks across the river are similar to those seen ahead on the eastern side of the Redford Basin.

Sierra Rica is the high mountain on the right horizon with Sierra el Mulato at mid-horizon and Sierra el Ranchito on the right. The Rio Grande on the right flows down the lowest block in the area.

Photographed at Mile 55.9

## The Presidio Basin

Just beyond Terneros Creek at Mile 58.2 the highway enters the Presidio Basin, 30 miles wide and 50 miles long. Underlying it is a graben, one of the largest and deepest in Texas, measuring 36 miles north to south by 18 miles east to west. The graben developed as the rift segment opened up, beginning at 25 Ma. The amount of extension, i.e. the amount it opened up, has been calculated at between one and two miles.

The graben is asymmetrical; movement along the western boundary fault was at least 10,000 feet; the eastern boundary is made up of multiple faults with total movement of about 4,300 feet. Fill is more fine-grained in the center of the graben than towards the margins where it becomes coarser and more angular.

I will discuss the Presidio Basin further in the next chapter.

### Bofecillos Peak

**Right** Bofecillos Peak (5,002 feet), built up of porphyritic basalt lava.

Photographed looking back from the rise at Mile 54.3 among banks and mounds of graben fill in the Redford Basin.

## Fort Leaton

The Fort Leaton State Park is on the left at Mile 62.3. The site and a surrounding 23.4 acres were acquired by the State of Texas in 1967 and opened to the public in 1978.

Ben Leaton, a Chihuahua Trail freighter and the earliest Anglo settler in the area, bought an old Spanish Fort, El Fortin de San Jose, in 1848, just after the Treaty of Guadalupe Hidalgo ended the Mexican-American War (1846-8) and formalized the Texas-Mexican border. The fort was founded in 1773, abandoned in 1810, and occupied as a private residence from 1830 by Juan Bustillo until purchased from him by Leaton.

Leaton rebuilt and expanded the fort making it an L-shaped building 200 feet by 140 feet with crenellated walls and an interior stockade for animals. He farmed along the river producing wheat, vegetables and feedstuff for his cattle and set up a trading post.

The fort was a welcome stopover for several of the early expeditions which were sent out from San Antonio to find an army supply route to El Paso, including the Hays expedition in 1848 and the Whiting expedition in 1849.

**Fort Leaton**

**Left** Fort Leaton was partially restored in 1934-5 and acquired by the state in 1967.

Photographed from Mile 62.3.

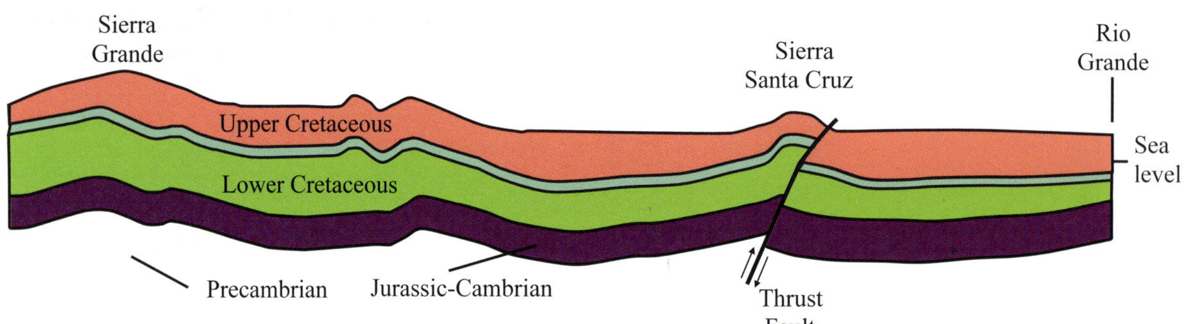


## Sierra La Santa Cruz

**Above** Sierra La Santa Cruz, a folded and faulted ridge of Cretaceous limestone. The fault, a thrust fault seen in the geological map about 3 miles south of Presidio, runs in front of the ridge and has brought up folded strata about 3,200 feet.

Mexican authorities in Ojinaga accused him of selling guns to Comanches in exchange for goods stolen in Mexico and said that he survived because he was useful to the Comanches. He died of yellow fever in 1851.

### Presidio

At about Mile 63.7, the highway enters Presidio, the largest town in Presidio County with a population given in the 2000 Census as 4,167. Although the area has supported a sizeable farming population along the Rio Grande and the Rio Conchos, which joins the former a short distance north of Presidio, the town of Presidio is of recent vintage. Spanish colonial administrations established garrisons at various times on the Mexican side of the river, but from the seventeenth century onwards, as the Comanches adopted to the horse and spread south into Texas, displaced Apaches and other tribes kept the area under constant pressure, and limited settlement.

A few brave souls ventured into farming along the river after the Treaty of Guadalupe Hidalgo, including John Spencer who had gained experience on the San Antonio-Chihuahua City trail as a freighter, a tough business in which the survivors knew how to defend themselves, but it was not until the Comanches were established in reservations in 1867 that Anglos began to settle in the area in small numbers and even by 1925, the town had a population of only 25. Today, Presidio has a flourishing cross-border trade with Ojinaga (population 35,000) across the river.

One much anticipated project that was expected to boost the economy was not successful. At Mile 64.6, on the eastern side of town, the highway crosses the South Orient Railroad. This line, originally the Kansas City, Mexico and Orient Railway and later owned by the Santa Fe railroad, was built to run the 1,600 miles between Kansas City, Missouri, and the port of Topolobampo on the west coast of Mexico. It was completed to Presidio in 1930 sharing tracks with what is now the Union Pacific line from Alpine to Paisano Pass. It then follows Alamito Creek almost to Presidio.

Although the route brings Kansas City 400 miles closer to the coast than Los Angeles, the line has never been much used. One problem is that its rails are lightweight and unable to handle standard freight cars. In 2000, the Texas Department of Transportation purchased the line from its previous owners and leased it for 40 years to the Mexican-owned company Texas Pacifico Transportation Ltd, which is said to be bringing the track back into service.

Presidio has a number of restaurants, gas stations and convenience stores along O'Reilly Street. Turn right on Erma Street and then left on Fourth Street to follow FR 170 to the junction with US 67 at Mile 66.7. Turn right again for Marfa.

**Sierra Rica from Presidio**

The great dome of Sierra Rica in Mexico (7,500 feet) is on the skyline in this photograph taken from Mile 64.8. The Rio Grande is in mid-picture.

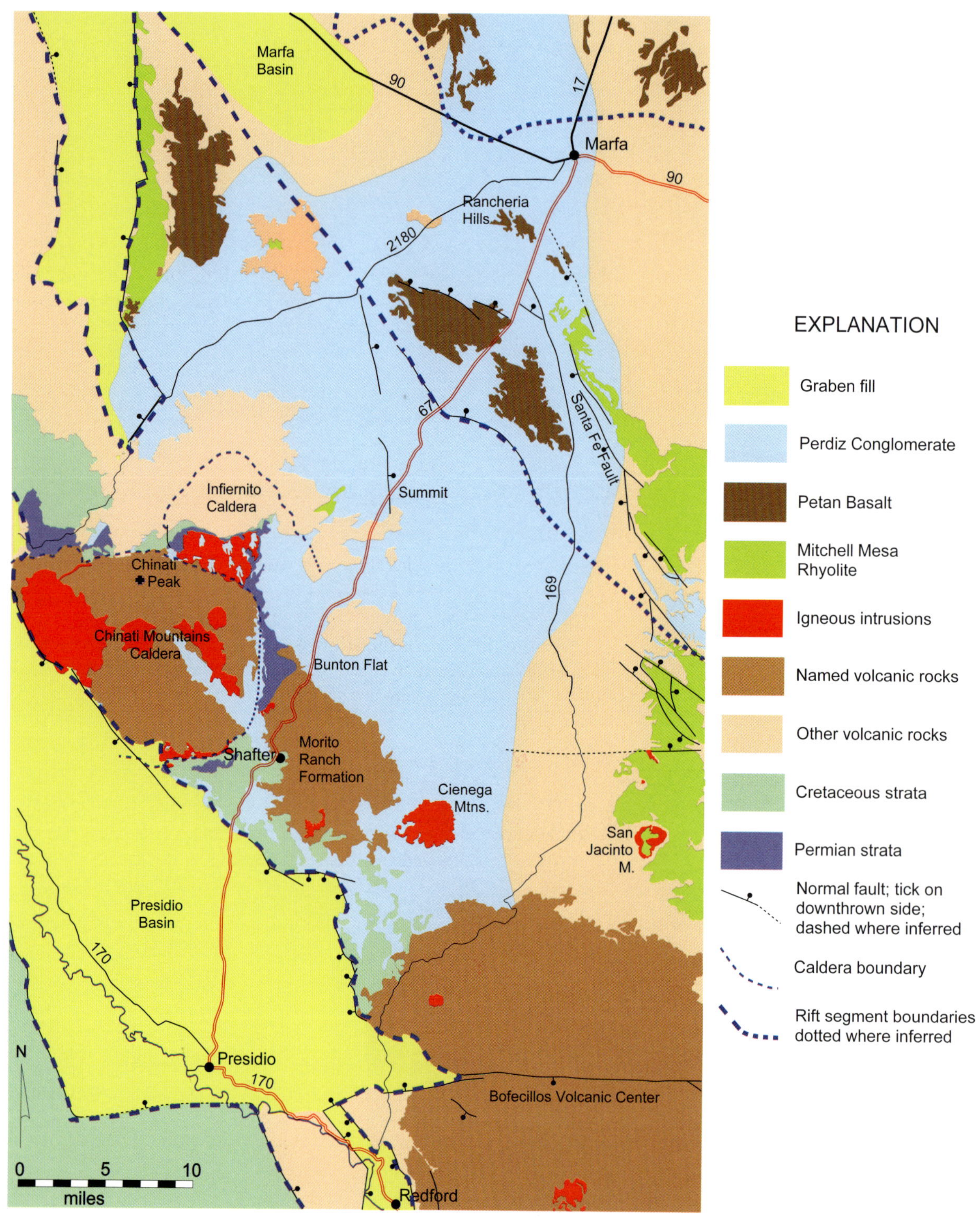


**Geology: Presidio to Marfa**

# 4 Presidio to Marfa

The Presidio-Marfa road, 58.1 miles long, begins at the junction of US 67 and FR 170 on the west side of Presidio, about one mile north of the bridge over the Rio Grande to Ojinaga. Mileages along the route are measured from there.

The geology is shown on the geological map opposite. The route begins in the Presidio Basin, crosses a saddle in a Cretaceous limestone ridge and descends through limestone road cuts into the valley of Cibolo Creek and the old mining district of Shafter. Immediately after Shafter the road climbs through multiple lava flows and tuff intervals of the Morita Ranch Formation on to Bunton Flat.

Beyond Bunton Flat, the road ascends rolling hills of Perdiz Conglomerate and lava flows to a summit almost 3,000 feet above Presidio. From there the road descends to the Marfa Basin, underlain by grabens of the Salt Basin rift segment. The segment's approximate boundaries are shown by the blue lines on the map, dashed where well-known, dotted where uncertain.

## The Satellite View

The satellite map on the next page gives a slightly larger view of the area. The graben fill of the Presidio Basin is dark gray, especially on the U.S. side of the river, with green irrigated fields along the Rio Conchos and Rio Grande.

The Rio Grande and Salt Basin rift segments are outlined in yellow. The Sierra Vieja, a range topped by the Bracks Rhyolite, a unit of similar age to the Crossen lava around Alpine, separates the two rift segments to the north. The boot-shaped structure to its right is a graben in the Salt Basin rift segment. To its left, a heavily-faulted area of volcanic rocks and light-colored Cretaceous limestones descend steeply into the Rio Grande rift segment along the river.

Southwest of the boot, the Infiernito Caldera has a well defined circle of lavas and tuffs in its center. Below it, the Chinati Mountains Caldera is noticeable mostly for the dark volcanic rocks within it.

Along US 67, the dark rocks to the east of Shafter are in the Morita Ranch Formation, a thick series of rhyolite and basalt lavas. The very dark patch southeast of Shafter is probably basalt lava.

**Satellite Map of the Presidio - Marfa Area**

North of Shafter, the light colored area crossing the road is the alluvium-covered Bunton Flat. To its north are two more Morita Ranch lava outcrop areas on the right of the road, both quite circular with dark centers.

Just before the FM 169 junction, Petan Basalt outcrops cross the highway, and again at Rancheria Hills. The Santa Fe Fault is very distinctive just to the right of Highway 169. The Marfa Basin itself is light greenish gray rather than silvery gray like the Presidio Basin, probably because it is grassier.

### The Presidio Basin

Leaving Presidio at the junction of FR 170 and US 67 west of town, the Marfa highway climbs gradually up the Presidio Basin, through desert terrain where only the creosote bush (*Larrea tridentate*) seems to survive. Some four terraces are cut into the slope, produced during periods when the river was halted in its down-cutting. The first begins at Mile 1.6 and continues past the airport. Occasional banks of graben fill appear along the highway and in road cuts. The most prominent bank, almost 300 feet high, runs to the right of the road for about 2½ miles from Presidio north (see photograph on page 134). Named on the USGS topographic map as Cibolo Hill (2,981 feet), it is known locally as the Desert Hills.

A second terrace begins at Mile 6.6. At Mile 15.1, the uppermost terrace flanks the highway at an elevation of 3,960 feet, 1,400 feet above Presidio (photograph on page 138). Farther down river, in Madera Canyon for example, graben fill is found at elevations as high as 4,600 feet and at nearly the same height on Burro Mesa in Big Bend National Park. If the fill had reached that height here, it would have buried the Livingston Hills (4,339 feet) on the left and would have reached as far as the hills above Shafter.

Picture then the scene after the volcanic eruptions ended, 27 Ma in the Bofecillos volcanic center and 32 Ma in the Chinati volcanic center. The low-lying area between the ridge above Shafter and Sierra Grande in Mexico filled with sand, silt and boulders brought in from the surrounding volcanic mountains by water draining into the area. Lakes would have developed behind obstructions downstream such as the Sierra del Carmen range at Boquillas Canyon. This range is now as much as 6,000 feet above sea level and was probably higher 27 million years ago. In fact, there is evidence that graben fill was deposited along the Sierra del Carmen in the last 10 million years by streams running north. This suggests that the saddle in the Sierra del Carmen through which the 1,500-foot deep Boquillas Canyon was cut did not develop until quite recently.

## Graben Fill in the Presidio Basin

**Above** Occasional banks of graben fill appear along the highway and in road cuts. The most prominent bank, almost 300 feet high, runs for about three miles to the right of the road north of Presidio. Named on the map as Cibolo Hill (2,981 feet), it is known locally as the Desert Hills. The great arch of Sierra Rica (7,250 feet) can be seen on the horizon behind it.

Photographed from Mile 3.3.

**Near Right** A close-up of graben fill, photographed three miles east of Presidio, at Mile 63.5 on the Study Butte-Presidio road. Generally, graben fill is more coarse-grained towards the edges of the Presidio Basin than in the interior of the basin.

**Far Right** More fine-grained graben fill photographed 7.3 miles east of Presidio, at Mile 59.5 on the Study Butte-Presidio road.

**The Presidio Basin**

The vast expanse of the Presidio Basin stretches out in front in this photograph taken at the pullout at Mile 10.7. The corrugated skyline ridge to the right is Sierra Grande, one of the major anticlines in the Chihuahua Tectonic Belt (see satellite photograph on page 132). Below it, the large populated area is Ojinaga, a town of some 35,000 people across the Rio Grande from Presidio.

The Rio Grande valley and Presidio are hidden behind the terrain in front of Ojinaga. The Desert Hills are in mid-photograph on page 136. To their right the prominent ridge is Sierra La Santa Cruz about five miles southeast of Presidio (page 126). Two small banks of graben fill are to the left of page 137.

## Livingston Hills

**Above** The Livingston Hills, of which Ross Peak (4,339 feet) is the highest, are just south of a crest in an anticline which has uplifted Cretaceous strata as you approach Shafter. The south side of the anticline is truncated by faulting along the edge of the Presidio Basin.

The flat in front of the camera is the fourth and uppermost terrace in the Presidio Basin, 1,400 feet above Presidio.

Photographed from Mile 15.9.

## Cretaceous Road Cut above Shafter

**Right** The north limb of the anticline is well exposed as you climb over the high point in the anticline at around Mile 17.0 and descend towards Marfa. In this photograph from Mile 17.4, Cretaceous strata in the east or right road cut are dipping at a low angle towards Shafter.

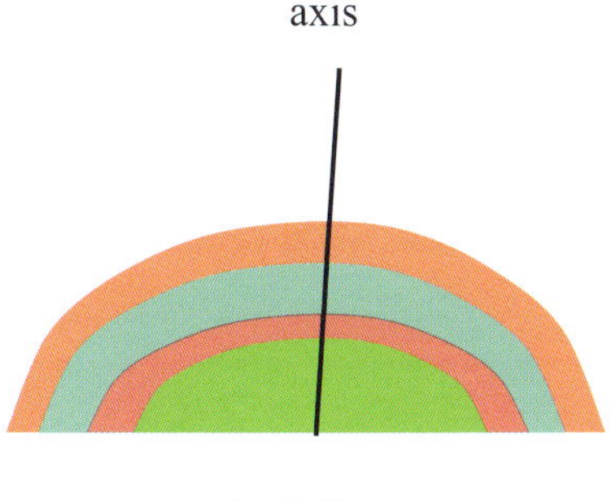


Anticline

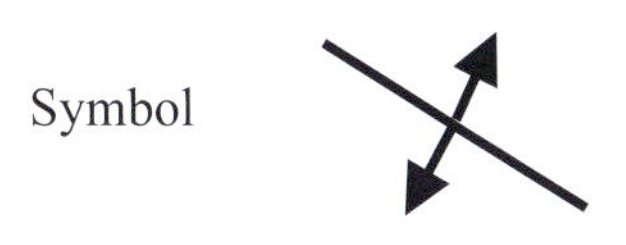


We know that the Rio Grande is about two million years old from examining sediments upstream in New Mexico. The age of the Rio Conchos, which today provides 65 per cent of the water in the Rio Grande at Presidio, is unknown. However, it probably developed before the Rio Grande because it rises only 200 miles away near Creel in Chihuahua State at 6,000 feet above sea level, whereas the Rio Grande does not reach this elevation until 550 miles upstream.

Water draining off the Sierra Madre Occidental would have coursed across basin fill to the Sierra Grande, found a way through that range and then flowed southeast to the Sierra del Carmen and turned north. As the Rio Grande rift segment deepened, the surface above the segment was lowered and the valley as we know it today began to form. Eventually faulting lowered the surface at Boquillas to the point where water poured over it and began creating Boquillas Canyon. From then on, the river level upstream was gradually lowered and other canyons began to develop.

## The Chinati Mountains from the South

This photograph of the Chinati Mountains from Mile 12.3 gives an oblique view of the front range. The pyramid on the right is Cerro Orona (6,305 feet). Most of the other summits rise up to about 6,000 feet elevation, 3,500 feet above Presidio. All are composed of lava flows of the Chinati Mountains Group.

Note the level banks of graben fill in the foreground.

## The Shafter Mineral Area

Excellent exposures of Cretaceous limestones can be seen along the highway from the Livingston Hills, 1,600 feet above Presidio, to Shafter. The limestones belong to the Cretaceous Shafter and Del Carmen Formations (see strata table on page 142) where about 1,600 feet of Lower Cretaceous strata are exposed along a ridge running northwest-southeast. To the northwest the ridge disappears under Chinati volcanic rocks; to the southeast, it dies away in the 12 miles to Alamito Creek valley. Borehole data indicates that Cretaceous and underlying strata have been raised as much as 2,500 feet along this ridge which is probably one of many formed during the Laramide Orogeny by compression from the west.

Silver, lead, copper, zinc and uranium minerals occur along a trend on the north flank of the ridge, as shown in the map on page 144. Although it is highly unlikely that the Spanish colonists of North America did not find silver ore at Shafter (they were, after all, the

most experienced silver miners in the world at the time), there is no evidence of mining in the area before John Spencer, a freighter who moved across the Rio Grande in 1854 to farm north of Presidio, found silver ore there in 1883. He showed a sample to William Shafter, commander of the First Infantry Regiment at Fort Davis, who had it assayed. When assay values proved encouraging, Shafter and fellow officers Lieutenants Bullis and Wilhelmi bought nine sections of land surrounding the prospect from the state and made Spencer a partner in the project.

Lacking mining expertise, Shafter, Bullis and Spencer first leased and then sold their interests to San Franciscan mining investors who set up the Presidio Mining Company in 1883. The three received 5,000 shares of company stock and $1,600 cash. Bullis, whose two sections had the best ore, refused to sell and slowed down development with litigation. However, in 1887 the company prevailed and began full-scale production at the Presidio Mine with an additional 300 workers.

### Strata around Shafter

| Period/Epoch | Formation | Age m.y. | Description |
|---|---|---|---|
| Oligocene | Chinati Mountains Group | 39.1 | Upper rhyolite, coarse grained, green more than 500 ft. thick.<br>Upper trachyte, olivine-augite trachyandesite, many thick flows, aphanitic to very fine grained, porphyritic, reddish brown to grayish black, 930 ft. thick.<br>Lower rhyolite, many thick flows, fine grained, porphyritic, light brownish black, 500 ft. thick.<br>Middle trachyte, many flows, very fine grained, porphyritic; medium to dark gray, light to dark olive gray and dark greenish gray; weathers reddish; 1,000 ft. thick.<br>Lower trachyte and tuff, very fine grained, amygdaloidal, porphyritic, dark reddish, interbedded with conglomeratic tuff, fine to coarse grained, greenish, 500 ft. thick.<br>Conglomerate; boulders, cobbles and pebbles in matrix of sand, silt and clay, all derived from Cretaceous rocks; 100 ft. thick.<br>Total formation thickness 3,500 ft. |
| | Morita Ranch | | Olivine basalt, grayish black, weathers to light to dark reddish brown, up to 250 ft. thick.<br>Ash-flow tuff, pale orange through grayish orange pink to reddish purple, weathers to light brown to very dark reddish brown, forms bluffs, up to 75 ft. thick.<br>Rhyolite, flow-banded and spherulitic, moderate red to dark yellowish brown, up to 300 ft. thick.<br>Basalt porphry, many flows, medium gray to black, weathers light to medium gray, up to 200 ft. thick. |
| Cretaceous | Del Carmen | | Limestone, microgranular to fine-grained, chert masses and beds up to 10 in. thick and 10 ft. long, gray, weathers shades of dark brown, yellowish brown and pinkish brown; 300-350 ft. thick. |
| | Shafter | ~125 | Limestone and sandstone, nodular to regularly bedded, alternately hard and marly forming a series of cuestas, sandstone predominant in basal 50 ft. and near middle of formation, fine to medium grained, thin to medium bedded, crossbedded, ripple marked, weathers dark reddish brown; marine fossils include *Exogyra texana* in upper 165 ft., *Orbitolina* in lower 700 ft., various other marine megafossils and petrified wood in some sandstone beds; 900 ft. thick. |
| | Presidio | ~140 | Conglomerate, limestone, sandstone and siltstone; basal pebble- to boulder-conglomerate, 30-80 ft. thick, grades upwards to slightly indurated pebbly sandstone and siltstone, followed by a succession of alternating impure carbonate rocks, calcereous sandstone, and siltstone; carbonate rocks weather yellowish gray, other rocks medium to dark gray weathering yellowish and reddish brown; *Orbitolina texana* common in upper half; about 400 ft. thick. |
| Permian | Mina Grande | ~260 | Limestone, hard, massive, gray to yellow to yellowish brown, up to 400 ft. thick. |
| | Ross Mine | ~270 | Alternating shale and sandstone, massive, black, weathers orange; limestone, massive, light to medium gray; sandstone, medium-grained; shale, silty, laminated, black; 727 ft. thick. |

The Presidio Mine was the largest silver mine in Texas between 1883 and 1942, with total recorded production of 2.306 million tons of ore containing 35.153 million ounces of silver at an average grade of 15.24 ounces silver per ton. Some 14 other prospects were explored along the ridge, although with little success.

The mine closed periodically during the 1920s as the price of silver fluctuated and was sold to the American Mining Company (now part of Cyprus Amax Mining Company) in 1928. The mine continued operating until 1942, when wartime manpower shortages and lower ore grades forced its closure. The rails underground were taken up and sold for wartime steel. Otherwise, the workings look very much as they did in 1942, having been preserved by the dry climate.

The Presidio Mine was purchased by Gold Fields Mining Company during an increase in silver prices when the Hunt brothers of Dallas attempted to corner the silver market in the 1970s. Gold Fields spent some $20 million dollars on the property, drilling 891 boreholes, sinking a 1,052-foot shaft and developing 5,100 feet of underground workings between 1977 and 1982.

When the silver price collapsed in 1984, however, Gold Fields sold the property to a small exploration company who resold it to the present owners, Silver Standard Resources, Inc. in 2000. Silver Standard, a publicly traded company on the Toronto and NASDAQ stock exchanges, which owns a number of silver prospects and producing properties in North and South America, calculates the reserves at 2.09 million tons of ore containing 10.06 ounces of silver per ton.

Silver Standard brought an 800 ton-per-day mill to the property in 2003 and has received the necessary permits for mining on the site. The company's operating plan proposes to mine 324,000 tons of ore per year for seven years, producing an average of 2.6 million ounces of silver per year to begin operations when circumstances permit.

### Geology of the Shafter Mineralization

The mineral deposits are found in the upper layer of Permian strata which consist of beds of limestone, clays and broken up rock called *breccia*, capped by a hard fine yellow limestone bed. During the 100 million years between the end of the Permian and the beginning of the Cretaceous, the area was on land, and the Permian landscape weathered with small caverns and sinkholes forming at the top of the formation (*karst* topography) before being overlain by Cretaceous strata of clay, limestone and conglomerate.

The Shafter silver deposits, *manto* deposits, are common in northern Mexico, New Mexico and Arizona. They were created by *hydrothermal* (hot water) solutions left over from magma as it

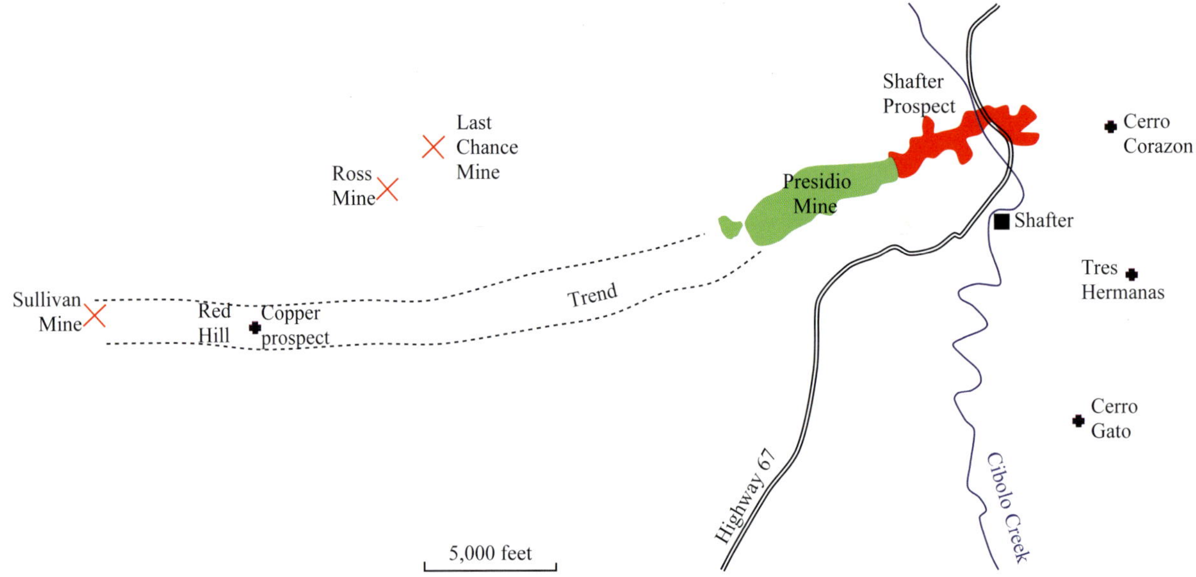


## The Shafter Mineral Area

The Shafter mineral area extends along an east to west trend for nearly two miles. The mined-out section of the old mine is shown in green, the prospect mining section in red. Some 14 smaller mines and prospects operated along the trend at various times, three of which are shown on the map; none produced ore in any quantity.

The Red Hill copper prospect was sampled by the Duval Company in 1970 and requires drilling to evaluate it. It is under option to Silver Standard Resources, Inc., the owner of the Shafter Mine.

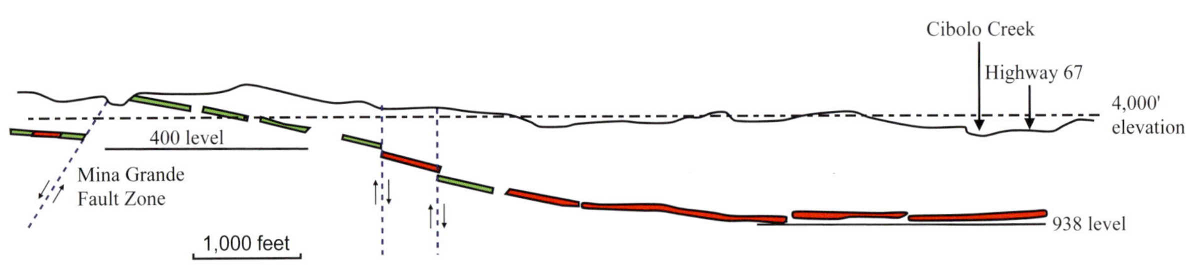


## Cross-section through Shafter Prospect

This cross section, taken from Silver Standard published material, shows the old workings in green and the prospective workings in red. Note that the workings rise nearly 1,000 feet to the left as strata climb the ridge. Faults are shown as dashed blue lines.

crystallized. The solutions followed permeable pathways in bedding, caverns, sinkholes and fracture zones. The elements deposited, silver, gold, lead, zinc, mercury and sulfur did not easily combine with the major constituents of the magma such as silicon, sodium, potassium and aluminum, and so became concentrated in the residual hot fluids. Silver, lead and zinc combined with sulfur to produce the main ores at Shafter, argentite, galena and sphalerite, sulfides respectively of silver, lead and zinc, along with some native silver.

The cross-section opposite shows how the mineralization deepens from the surface on the west to almost 1,000 feet below the surface at Cibolo Creek as it follows the contact between Permian and Cretaceous rocks.

The magma source of the mineralizing fluids is not known. There is no sign of heat around the mining operations so it could not have been in the immediate vicinity. The Red Hill copper prospect adjoins a quartz monzonite intrusion at the surface along the same trend. At one time this intrusion was thought the source but recent work has shown that it is 60 to 64 million years old, similar in age to copper-molybdenum deposits in Arizona and New Mexico, and so predates the Shafter silver deposits by at least 30 million years. Now the most likely source is thought to be intrusions that came in along the southern collapse zone of the Chinati Mountains Caldera.

Drilling logs indicate that there is a 110-foot zone on the eastern edge of the Red Hill intrusion which grades six percent copper. The prospect needs further exploration to determine whether it has commercial potential or not.

### Shafter

You can see some of the mining equipment on the right at Mile 18.5 as you go downhill to Shafter. The town grew up around the mine's mill, whose foundations can still be seen on the right at the base of the hill. A post office opened in the town in 1885 and by 1900, the town had two saloons, a dance hall, a school and a population of 110.

By the time the Presidio Mine closed, Shafter had a flourishing economy based on the two military bases in the county, Marfa Army Air Field and Fort D. A. Russell. In 1943 it had a population of 1,500 with twelve businesses serving the military population. After the bases closed, the population of Shafter declined, reaching twenty by 1949. The population was twenty-six in 2000.

Just beyond Shafter the road crosses Cibolo Creek at Mile 18.6 and immediately begins climbing through road cuts of the Morita Ranch Formation (see table on page 142), a series of rhyolite and basalt lava flows with interbedded tuffs. Their source is uncertain but

**Above** Panoramic view of Shafter today, Cerro Corazon on the left, Tres Hermanas to the right and Cerro Gato on the far right. Note Morita Ranch tuff cropping out on the flanks of Tres Hermanas.

Photographed from Mile 18.0.

**Below Left** The Presidio Mine, Shafter, in full operation, prior to 1942 when it closed down. The large building in the center of the photograph was a mill. It crushed ore taken from the mine and separated material containing silver minerals from waste. The silver concentrate, as it was called, was sent off for refining. Some of the waste can still be seen on the incline south of Shafter Ghost Town.

**Below** The old church in Shafter still holds Sunday services.

Photographed from Mile 18.4.

they may have come from the Infiernito Caldera and are older than the Chinati Mountains Group.

### Cibolo Creek and Cienega Ranches

At the top of the rise at Mile 24.7, the road comes on to Bunton Flat, overlying a graben faulted on its southern side. The historic ranches of Milton Faver are ahead, Cienega Ranch on the right and Cibolo Creek Ranch on the left. Faver, the first cattle baron in Presidio County, controlled these ranches in the nineteenth century along with the La Morita Ranch, five miles southeast of Shafter.

Like Ben Leaton and John Spencer, Milton Faver had been a freighter on the Santa Fe and Chihuahua trails before operating a general store in Ojinaga. He moved to the Chinati Mountains in 1857 and began acquiring small parcels of land around the Cienega, Big Spring Cibolo, Cibolo, and La Morita springs, and built up a cattle herd. His biggest customer was the U.S. Army at Fort Davis and other camps.

Federal forces withdrew from the area at the outbreak of the Civil War in 1861 and left Faver without a market or protection from

## Morita Ranch Formation Terrain

Just beyond Shafter, on the far side of the bridge over Cibolo Creek at Mile 18.6, the road enters volcanic terrain. At roadside and on the hills to the right behind Shafter, the rocks belong to the Morita Ranch Formation. This formation consists of, from the bottom up, several basalt lava flows up to 250 feet thick, a 75-foot thick layer of ash-flow tuff, a succession of rhyolite lava flows up to 300 feet thick and many flows of basalt porphyry up to 200 feet thick. The highway winds up among Morito Ranch volcanic outcrops past Cerro Corazon (5,000 feet) on the right. An excellent exposure of a boulder conglomerate can be seen in a road cut on the left at Mile 20.5, probably quite young material.

**Left** The stream in the foreground of the photograph has carved a deep valley in volcanic rocks of the Morito Ranch Formation.

Photographed from Mile 21.0.

**Below** In this photograph taken from Mile 23.1, a Cretaceous mesa is in front of a highly tilted block. These blocks are probably near the perimeter of the Chinati Mountains Caldera in which the surface collapsed, creating jumbled blocks along its perimeter.

Indian raids. His cattle were mostly stolen but he and his Mexican wife survived, thanks to the forts he had built on the Cibolo Creek and Cienega ranches. After the Civil War, Faver rebuilt his cattle herd and resumed selling beef to the Army, eventually owning more than 10,000 cattle. He built a third fort on the La Morita ranch after the ranch foreman, his brother-in-law, and his family were abducted and killed in an Indian raid in 1875.

The Faver family lived at the Cibolo Creek fort with vegetable gardens and a peach orchard well watered by the Big Springs Cibolo. Many early travelers enjoyed the Faver hospitality, especially his notable peach brandy. Milton Faver died of natural causes in 1889.

Since 1988, all three ranches have been owned by John Poindexter, a Houston industrialist, who has done much to improve the properties, restoring natural grasses, eliminating invasive plant species, and improving many thousand acres of eroded landscape. The ranches continue to be working ranches, stocked with Longhorns. The three old forts have been renovated and are jointly

**Chinati Peak**

**Right** A close up view of Chinati Peak from Mile 25.6. The mountain is built up of many rhyolite and trachyte lava flows of the Chinati Mountains Group. Some large igneous intrusions flank the summit to the west, south and northeast.

operated as a resort. The largest, at Cibolo Creek, has an adjoining hacienda with 21 guest rooms; the Cienega fort has five guest rooms with another five in an adjoining hacienda; while remote La Morita has a one-bedroom guest cottage.

### The Chinati Mountains Caldera

At Mile 25.6, the beautiful Chinati Mountains amphitheater opens up at 9 o'clock. The amphitheater overlies the Chinati Mountains Caldera, the largest in West Texas, 18 miles long and 12 miles wide. It is overlain by a series of rhyolite and trachyte lava flows and tuffs over 3,000 feet thick, and several large intrusions. The volcanic rocks date from 32 to 31 Ma and are thickest on Chinati Peak (7,730 feet) whose enormous dome dominates the landscape for many miles around.

The caldera collapsed before the visible volcanic units emerged, and although the caldera wall is visible in places, no one knows how deep it goes. The caldera wall runs quite close to the highway at Shafter, and in fact some of the faulting along which silver minerals arrived probably derived from the caldera collapse.

**Cienega Mountains**

**Left** The Cienega Mountains are formed from a very large rhyolitic volcanic dome 4 miles in diameter and rising 2,000 feet above the plain.

Photographed from Mile 25.5.

The caldera is also thought to have been the source of the Mitchell Mesa Rhyolite, a welded tuff found over an enormous area from near Alpine to northern Chihuahua. You can see it along the river near Redford and on distant escarpments along Green Valley. The rhyolite came from one of the great eruptions of the area at around 32 Ma. It spread over 6,000 square miles to a depth of 6 to 600 feet, in all a volume of about 250 square miles. If it came from the Chinati Mountains Caldera which has a surface area of about 200 square miles, it would have created a caldera 6,600 feet deep.

The panorama above was photographed from Mile 27.5. The range stretches from Chinati Peak (7,730 feet), the third highest peak in Texas, on the right. Cerro Alto (4,885 feet), the mesa in the right foreground is capped by a flat slab of Morito Ranch Formation lava.

### Perdiz Conglomerate

From about Mile 31 to Marfa, the highway runs on a formation called the Perdiz Conglomerate, a series of loosely cemented siltstones, sandstones and conglomerates that originated in fans of alluvial material eroded off the Chinati Mountains and other high points in the area. At first some lava beds can be seen in the conglomerate, at Mile 33.7, for example, where a lens of porphyritic orange-brown lava, 45 yards long and up to 4 feet thick is embedded in Perdiz siltstone in the right road cut.

### Chinati Mountains Caldera

**Above** The long range of the Chinati Mountains comes into view on the left as you come on to Bunton Flat. The range stretches from Chinati Peak (7,730 feet), the third highest peak in Texas, on the right. Cerro Alto (4,885 feet), the mesa in the right foreground is capped by a flat slab of Morito Ranch Formation lava.

Photographed from Mile 27.5.

**Right** This pyramid (5,140 feet) is prominent on the horizon to the left of the highway from about Mile 26 onwards.

The hill is a small remnant of Mitchell Mesa Rhyolite from the Chinati Mountains Caldera overlying ash-flow tuff also from the caldera, which has been protected from erosion by the rhyolite.

Photographed from Mile 30.2.

### Perdiz Conglomerate

**Above** Two views of the Perdiz Conglomerate: on the left near the base of the formation, pebbly sandstone and siltstone beds of the Perdiz Conglomerate at Mile 39.3; on the right at the highest point on the highway at Mile 36.3, more a true conglomerate.

The formation crops out in an area 40 by 25 miles centered on US 67. It was created by erosion of the Chinati Mountains and other high areas to the west and accumulated as a set of alluvial fans which amalgamated into the present formation.

The Perdiz Formation is thought to be about 29 million years old. See table on page 16 for definitions of conglomerate, sandstone and siltstone.

**Above Right** Typical Perdiz Conglomerate terrain, mildly dissected landscape characterized by gently rolling hills and broad stream valleys.

Photographed from Mile 30.4.

The last volcanic outcrop is the very prominent sharp pyramid-shaped hill (5,140 feet) on the left of the road which is visible on the horizon from about Mile 26 onwards. It is capped by a small Mitchell Mesa Rhyolite remnant overlying tuff beds from the Chinati Mountains Caldera which have been protected from erosion by the rhyolite (photograph on page 153). Geologists call this type of outcrop an outlier, one that is surrounded by older rocks.

The Perdiz formation is exposed intermittently in road cuts as you climb up 800 feet to the summit of a ridge at Mile 36.3. The formation changes during the ascent from siltstone (much of it from volcanic ash) to siltstone with pebble beds to true conglomerate at the top, full of rounded boulders.

The pullout at the summit is the high point on the Presidio-Marfa highway, 5,420 feet above sea level, and is a wonderful viewpoint over the region. Taking the highway ahead as 12 o'clock, the Chinati Mountains are from 6 to 7 o'clock, Sierra Rica in Mexico at 5 o'clock, the flat-topped Santiago Peak at 3 o'clock, Cathedral and Goat

## Cathedral and Goat Mountains

Looking to the east (right) of the highway, Cathedral Mountain (6,800 feet) on the left and Goat Mountain (6,608 feet) on the right are eroded remnants of volcanic rocks that once covered the entire area up to a height of nearly 7,000 feet. They owe their existence to the presence on their summits of a thick Petan Basalt lava bed that held back erosion.

The basalt lava caps the ledge on the left of Cathedral Mountain and on the left side of Goat Mountain. The ledge on the right of Goat Mountain is formed by Mitchell Mesa

Rhyolite. The spire on Cathedral Mountain is a block that has been faulted by an intrusion pushing up from below. For another view of Cathedral Mountain see page 30.

Light-colored tuff in the escarpment of the Santa Fe Fault can be seen dimly in the middle distance on the right.

Photographed from Mile 38.4.

## The Marfa Basin

Looking north from the same viewpoint as that of the photograph on the previous page, the generally level appearance of the Marfa Basin is striking. The Puertacitas Mountains are on the skyline on the left. One of the two Haystack intrusions is on the right skyline; the other Haystack intrusion is hidden behind it.

The low mesas in front of the Haystacks and in mid-photograph on the far left are capped by Petan Basalt.

## Basalt Feeder Dike in Tuff

**Above** In this photograph, taken from Mile 45.9, a bed of white tuff capped by Petan Basalt is exposed in the west road cut. Note the thin basalt dike running from road level to the basalt cap. We generally think of dikes as great thick bodies of rock, as they usually are when composed of viscous rhyolite. Basalt is much less viscous and can flow through narrow fissures, as this example shows.

## Rancheria Hills

**Right** Low hills capped by Petan Basalt on the left of the highway, photographed from Mile 51.7. Light gray tuff crops out in the shadows below the basalt beds.

Mountains at 1 o'clock, Paisano Peak at 12 o'clock, the Puertacitas Mountains at 11 o'clock and the Davis Mountains at 10 o'clock.

From the summit, the highway descends 700 feet through mildly dissected rolling terrain to the flat Marfa Basin at Mile 41.2, with intermittent exposures of Perdiz conglomerates, sandstones and siltstones along the way.

### The Marfa Basin

The Marfa Basin is a rolling plain under the final 16 miles of the Presidio-Marfa highway with mudstones of the Perdiz Conglomerate cropping out in road cuts as far as Mile 45.0.

Under the plain are grabens of the Salt Basin rift segment. The exact southern boundary of the segment is unknown but a well drilled on the right shortly after Perdiz Creek at Mile 41.8 found the base of the volcanic rocks at 617 feet above sea level, much lower than in the previous well along the highway, at Mile 28, where the base was at 4,190 feet above sea level.

Clearly, the first well was drilled in a graben, and the likely southern boundary of this graben is the point at which the road comes down on to the flat at Mile 41.2, just before Perdiz Creek, also the

point where the geological map shows a fault crossing the highway with down-throw to the north.

The rift segment is made up of grabens interspersed with horsts. In several places, Petan Basalt beds crop out on the surface, indicating the presence of horsts. One well drilled near basalt lava about Mile 48.7 found the base of the volcanic rocks at 4,150 feet elevation, so the well must have been drilled into a horst. On the other hand, several wells between here and Marfa found the base at 800 feet above sea level or lower and must have been drilled into grabens.

The Salt Basin rift segment seems to deepen northwest of Marfa. One well, about 10 miles northwest along US 90 found the base of the volcanic rocks at 100 feet below sea level. Another deep well near Valentine, 35 miles northwest of Marfa, intersected the base at 2,094 feet below sea level.

From about Mile 45.8 to 48.3, the highway cuts through basalt lava. These lava flows belong to the Petan Basalt formation which, from about 23 to 18 Ma erupted on the landscape through graben faults, presumably from the hot interior currents, and are found widely in the Big Bend. Some of the road cuts, such as one in the photograph on page 160, have interbedded white tuff lenses. This material is similar to the Perdiz Conglomerate in that it formed from sediments eroded off tuff exposures after the end of the main volcanic activity.

## Marfa

Marfa, like Alpine and many other towns in West Texas, began as a watering point on the Galveston, Harrisburg and San Antonio Railway in 1883. Railroad steam engines of that era had to have water every thirty miles or so. The name supposedly came from a character in a Russian novel being read at the time by the wife of one of the railroad company's executives. Marfa is the Russian for Martha.

The town quickly became a distribution point for its area and its standing was enhanced when Presidio County moved its headquarters there from Fort Davis in 1885. The elegant courthouse was built at a cost of $60,000 in 1886. By 1900, the population was 900.

Marfa had a military presence for many years, beginning in 1911 when a cavalry company was stationed south of town. Activity increased during the Mexican revolution when Camp Marfa, as it was called, became headquarters for the Big Bend district and biplanes reconnoitered the border from canvas hangers south of town.

One of the very best books about early West Texas, *Chronicles of the Big Bend* by W.D. Smithers, published by Texas A&M Press and still in print, has reminiscences and photographs of this period, including tales of driving mule teams on the two-day journey from Marfa to Presidio.

Camp Marfa changed its name to Fort D. A. Russell in 1930 and became a temporary prisoner of war camp during World War II. Although military bases closed at the end of the war, the Federal government again has a large presence through the Border Patrol which controls immigration in much of West Texas and Oklahoma from offices in Marfa.

Marfa has a growing art influence in the southwest thanks to the late Donald Judd, the sculptor and minimalist painter, who purchased the old fort in the late 1960s and transformed it into an art museum. The museum opened in 1986 and is run by the non-profit Chinati Foundation which also sponsors art and education programs throughout the year and runs an Art Festival in October each year, attended by over 2,000 visitors from many parts of the globe.

### Presidio County Courthouse

**Left** The Presidio County Courthouse in Marfa was built in 1885 at a cost of $60,000, an enormous sum for those days, to a design in the Second Empire style with Italianate flourishes. It was refurbished in 2001.

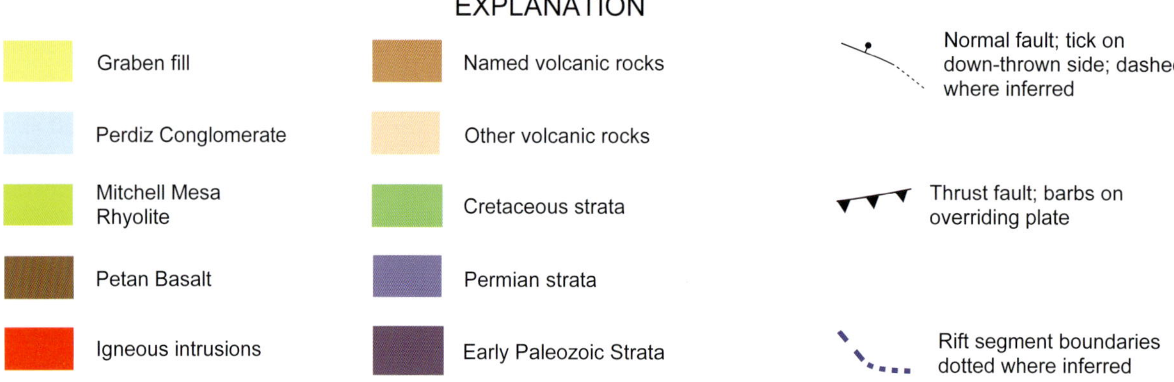

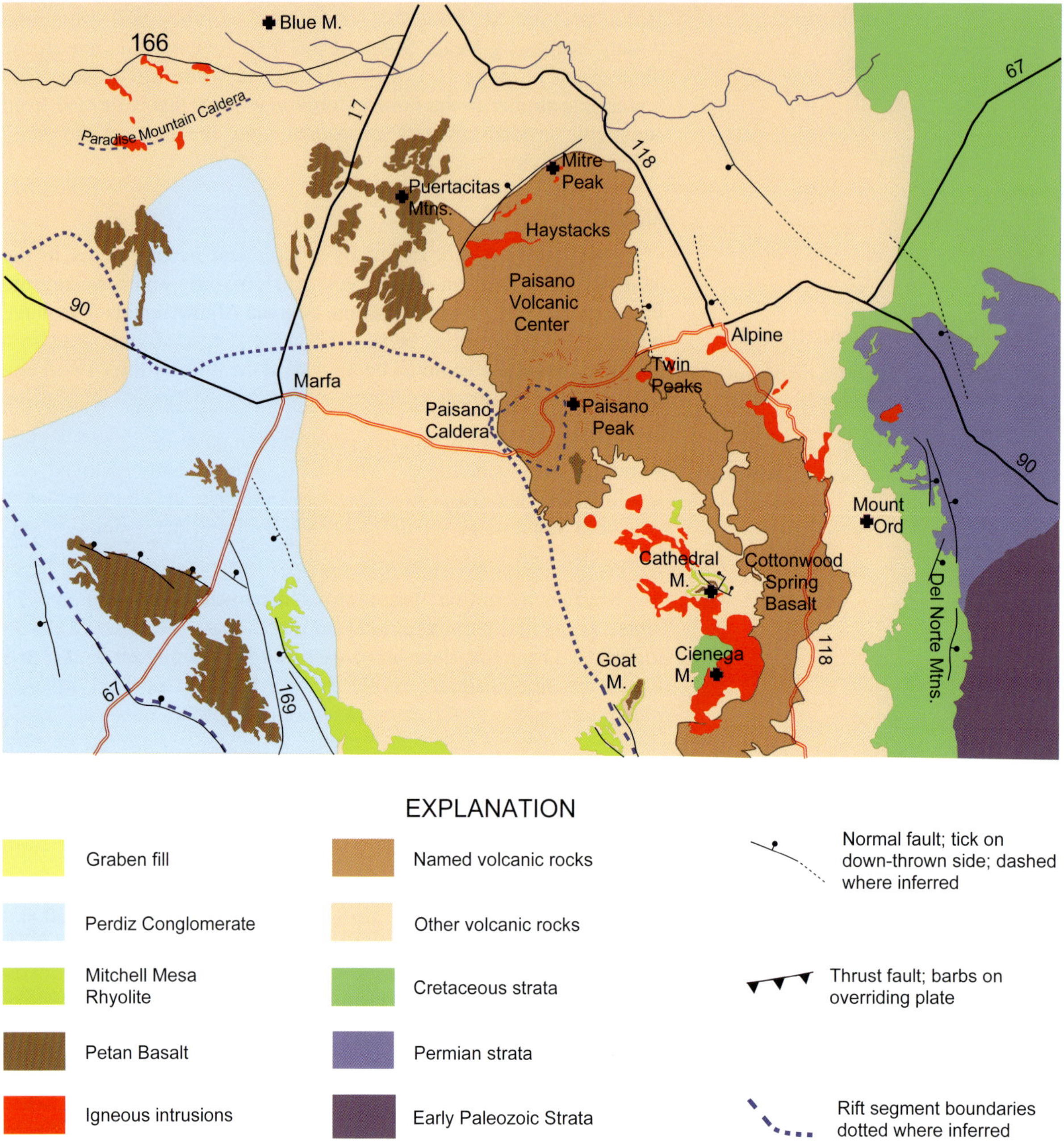

**Geology: Marfa to Alpine**

# 5 Marfa to Alpine

The Marfa-Alpine segment of River Road loop begins in the Marfa Basin, and drives through the Paisano Volcano and the Paisano Caldera before coming out into the Alpine Basin. This is the only opportunity in the Big Bend to see the results of a magma chamber collapse close up.

The segment is 26.9 miles long. Mileages are measured from the flashing red light at the intersection of US 90 and US 67.

The Marfa Lights Viewing Area on the right at Mile 8.6 was built to provide a site to view the lights, an unexplained natural phenomenon, and is a good point to stop and view the scenery. The lights are said to be visible every clear night between Marfa and Paisano Pass as one faces the Chinati Mountains. At times the lights appear colored as they twinkle in the distance; they move about, split apart, melt together, disappear, and reappear. Local residents have reported seeing the lights for well over a hundred years.

On aerial photographs, two runways of the Marfa Army Air Field show up clearly in front of the viewing area. The air field was used to train air crew during World War II, with as many as 30,000 men stationed there at any one time, sleeping under canvas. Pilots in training looked for the source of the elusive lights from the air, again with no success.

At Mile 12.3, the highway crosses the South Orient Railroad, which we last crossed on the outskirts of Presidio. The railroad from this point to Alpine shares the Union Pacific track that the highway has followed from Marfa.

### The Paisano Volcanic Center

The highway enters the Paisano Caldera just after the railway bridge. The cliffs ahead and to the right are made up of lavas of the Decie Formation produced by the Paisano Volcanic Center, which was active for about a million years around 36.3 Ma.

The Decie Formation is divided into five members (see table on page 166) and consists of rhyolite and quartz trachyte lavas with interbedded tuffs, in total up to 3,000 feet thick. The lavas and tuffs erupted from multiple fissures. The contents of the fissures solidified into dikes when the supply of magma was exhausted and the flow upwards ended. As many as a thousand dikes crop out in the Paisano

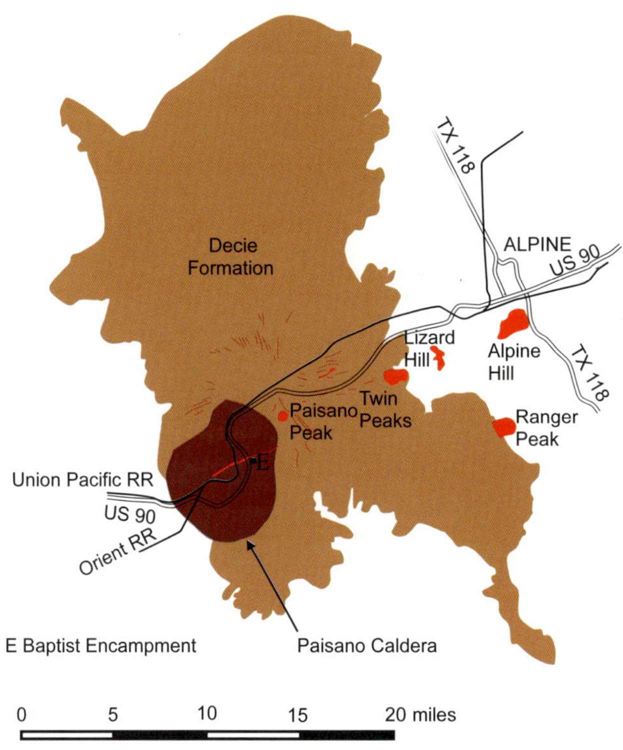


**Geology: The Decie Formation**

## Strata of the Decie Formation

The Decie Formation caps the cliffs to the west of Alpine. In geology, a formation is a body of rock identified by its physical characteristics and position in the geological time scale. A formation can be subdivided into members or grouped with other formations into a group.

The formation consists of rhyolite and quartz trachyte lavas with interbedded tuffs; source multiple fissures 10 miles west of Alpine; total thickness up to 3,000 ft. Five members have been identified:

*McIntyre Lava Member:* Quartz trachyte, trachyte, some rhyolite, minor agglomeratic tuff, conglomerate, and ash-flow tuff; up to 1,100 ft. thick.

*McIntyre Tuff Member:* Agglomeratic tuff and several small ash-flow sheets; 0-300 ft. thick, thinning towards Alpine.

*Morrow Lava Member:* Quartz trachyte, slightly porphyritic in places; up to 750 ft. thick.

*Morrow Tuff Member:* Yellow agglomeratic tuff interlayered with two minor quartz trachyte ash-flow sheets; up to 300 ft. thick, thinning towards Alpine.

*Rhyolite Member:* Flows, domes and plugs of rhyolite with minor volcanic breccia, bedded tuff and poorly welded ash-flow tuff; up to 600 ft. thick.

## The Haystacks

**Above** The Haystacks are twin trachyte intrusions, 6,895 and 6,670 feet high. They probably both came from the same source.

Since they are both intrusions, i.e. they solidified underground, we can assume that the volcanic blanket in this area was 7,000 feet above sea level when they formed and may have extended to Cathedral and Goat Mountains, both with summits about 6,800 feet above sea level today.

The rocks in the volcanic blanket were probably similar to those seen today on Cathedral and Goat Mountains where they were preserved by abnormally thick basalt lava beds on their summits.

### Chinati Peak

**Above** Chinati Peak photographed from the Marfa Lights Viewing Center at daybreak.

### Chaotic Terrain in the Paisano Caldera

**Right** An example of material in the Paisano Caldera, photographed at Mile 15.8. On the left, a red block of Morrow welded tuff is in a light gray pebbly lahar.

A *lahar* is a mudflow, consisting mainly of volcanic rocks and debris, on the flanks of a volcano.

volcanic center, some of which are shown in red on the map on page 166.

The Decie Formation overlies the Cottonwood Spring Basalt, a series of basalt lavas and interbedded tuffs found around and south of Alpine.

The caldera formed after eruption of the McIntyre tuff and before eruption of McIntyre lava. Strata older than the latter are very broken up, strata younger only slightly so. The amount of collapse into the caldera varies widely, from 1,000 feet west of Paisano Peak to negligible amounts in other places. The caldera boundary runs along the low ground between the railway line and the hills on the left skyline and crosses the road again six miles ahead. It then runs approximately parallel to the highway along the high hills behind the Paisano Baptist Encampment and returns to this point, passing about a half mile west of Paisano Peak which is not in the caldera.

A Texas Historical Commission plaque marks the summit of Paisano Pass (5,074 feet) at Mile 12.6. When the Southern Pacific Railroad built its track through here in 1882, the pass was said to be the highest point on the line. It is also claimed to be the highest point on Highway 90 between Florida and California.

Road cuts along the next four miles show how the caldera collapse created jumbled groups of rocks caught up in mud flows. During the volcanic era in west Texas the climate was wet and warm, which led to plentiful mud.

Just before the Paisano Baptist Encampment entrance at Mile 15.1, the thick dike exposed in a road cut and creating a ridge on the left is three miles long, one of the largest mapped in the volcano. The 500 foot high walls at 3 o'clock behind the encampment are on the eastern edge of the caldera where the collapse was around 1,000 feet.

The road cut on the right at Mile 15.3 shows the tremendous variety of rock types in the caldera. Beginning on the right, red-

**Right** At the top of the rise at Mile 19.3, welded ash-flow Morrow tuff on the right road cut includes a circular trachyte inclusion, 3 feet in diameter. No trachyte erupted in the volcano before the Morrow tuff, so this and other trachyte inclusions must have been brought up from the Earth's crust below the Decie strata.

The road cut runs along the edge of a quartz trachyte dike cutting the rhyolite. Striations in the gray wall show that the dike flowed at roughly 60 degrees up to the east.

brown welded tuff is followed by a section of very poorly welded tuff with red-brown blocks of volcanic breccia, then a large block of volcanic breccia between two sections of poorly welded yellow gray tuff 10 and 3 feet wide. Next, a patch of welded volcanic breccia in red brown tuff is followed by a fossil mudflow containing a red brown tuff block oriented at 45 degrees to the right (see photograph on page 169). The jumbled cut with its randomly oriented blocks suggests that the rocks resulted from a landslide into the caldera. Much of the cut overhangs the ground below and is best viewed from the left of the road.

At Mile 20.1, a block of yellowish-gray bedded Morrow tuff is faulted down against pale pink to dark red Paisano rhyolite. In places, the rhyolite is a very pretty spotted rock, light bluish gray on fresh surfaces, which was given the name *paisanite* by one of the early geologists to visit the area but is usually shattered and oxidized along this road. It turns yellowish when oxidized.

The high walls around the picnic area at Mile 20.7 are capped by McIntyre lava underlain by Morrow lavas and tuffs. A ridge across

**Left** A lahar between tuff blocks, photographed at Mile 16.9. The lahar is composed mainly of pebbles and boulders in a tuff debris matrix.

the highway from the picnic area is surmounted by the ragged crest of one of the Paisano volcanic feeder dikes (photograph on page 173), which continues parallel to the road for the next half mile. Others can be seen behind the picnic area.

Two openings separated by a ridge 750 feet high carry the highway and the railroad out of the cliffs at Mile 22.1. Horizontal quartz trachyte flows of Morrow lava cap the high ground on either side underlain by Morrow tuffs with Cottonwood Spring basalt cropping out at the base of the escarpments.

The exit from the cliffs at Mile 22.1 is one of the best viewpoints across the basin. The escarpment running north along the skyline at 10 o'clock is what geologists call a *fault escarpment*, caused by strata breaking along the line of the escarpment and dropping down on the right relative to strata on the left. The fault escarpment runs along the southwestern boundary of the Alpine Basin. Where the fault crosses the road, movement has been about 300 feet; at Sunny Glen farther north, movement has been 750 feet. Within the basin, alluvial sand and gravel up to 75 feet thick cover basalt bedrock.

**Right** This dike, photographed from the picnic area at Mile 21.7 runs parallel to the road for the next mile or two.

It formed in a fissure that brought magma up to the surface of the volcano, one of perhaps a thousand such fissures in the Paisano Volcano. When magma stopped flowing, the material in the fissure solidified and now appears as a dike, a long thin vertical slab of rock.

Looking to the left, the basin continues up to Polks Peak (5,304 feet), the hat-like butte 19 miles away at 10:30 o'clock. On the right, the cliffs continue round the south rim of the basin to Bullfrog Mountain on the horizon at 12:30 o'clock. Hancock Hill is at 12 o'clock.

The road descends a sloping erosional surface or pediment as it approaches town. Lizard Mountain is the rough quartz trachyte intrusion at 3 o'clock from Mile 23.5. A trachyte dome to its right was uplifted by an intrusion. The intrusions on the skyline are Ranger Peak (6,246 feet) 2½ miles away at 4 o'clock, and the double-headed quartz trachyte Twin Peaks (6,133 feet and 6,112 feet) at 5 o'clock.

The junction of US 90 and TX 118 South at Mile 26.3 brings us back to the point where we began this loop 230 miles ago.

**Left** The elegant rounded silhouette of Paisano Peak (6,085 feet) rises 1,050 feet above the highway, a nepheline syenite plug intruded into the Decie Rhyolite after volcanic activity had ended. The base of the intrusion, which can be seen on the opposite side of the peak, is about one-third of the way down the slope.

Photographed from Mile 19.9.

**Alpine at Nightfall**

One of the great pleasures of living in Alpine is watching the sun go down over the western hills at nightfall.

This photograph is taken from the same vantage point as the morning view on page 24, looking west across the Alpine Basin.

## Acknowledgments

*With thanks to*

Pat Dickerson for providing an outline of her latest work and her dissertation on the Tascotal Mesa Fault, reading the tectonic sections of the manuscript and providing a detailed critique that improved the manuscript enormously.

Pat Dasch and Martha MacLeod for proof-reading the manuscript

Scott Baldridge for reading the introduction to the Salt Basin Rift and the Sunken Block and for providing stimulating discussions on Western American tectonics over the last several years

Jack Burgess for correcting the Shafter Prospect description

Ken Barnes for providing information on the Big Bend Museum of Paleontology and Geology

Don Dowdey, Sul Ross Library, for checking the Publishers Cataloging in Publication Data

# Notes

**Page**

10 This map is a composite of the Fort Stockton, Marfa and Emory Peak-Presidio sheets, scale 1:250,000, the Tectonic Map of the Basin and Range Province of Trans-Pecos Texas, scale 1:500,000, and mapping by Christopher Henry of Big Bend Ranch State Park, all published by the Bureau of Economic Geology.

39 Recent work in the Rocky Mountains includes: McMillan, M.E., Heller, P.L. and S.L. Wing, 2006, History and causes of post-Laramide relief in the Rocky Mountain orogenic plateau, *GSA Bulletin*; March/April 2006; v. 118; no. 3/4; p. 393–405.

41 The depth of the Sunken Block is from Dickerson, P. W. and W.R. Muehlberger, 1994, Basins in the Big Bend segment of the Rio Grande rift, *in* Keller, G.R. and S.M. Cather, eds., Basins of the Rio Grande rift: Structures, stratigraphy, and tectonic setting: Geological Society of America Special Paper 291, p. 283-297.

65 Data on the Solitario comes from: Henry, C.D., Kunk M.J., Muehlberger,W.R. and W.C. McIntosh, 1997, Igneous evolution of a complex laccolith-caldera, The Solitario, Trans-Pecos Texas: Implications for calderas and subjacent plutons, GSA Bulletin, v.109, p 1036-54.

65 The date of the basalt sill in the Mesa de Anguila is from Muehlberger, W.R., 1989, Summary and structure of Big Bend National Park and vicinity *in* Muehlberger, W,R, and Dickerson, P.W., eds., Structure and Stratigraphy of Trans-Pecos Texas, Field Trip Guidebook T317, 28th. International Geological Congress, American Geophysical Union, Washington, D.C.

83 Information on the rift along the river came from: Henry, C.D., 1998, Geology of Big Bend Ranch State Park, Texas, Bureau of Economic Geology, Austin, Guidebook 27, 72 p.

and from: Henry, C.D., 1998, Basement-controlled transfer zones in an area of low-magnitude extension, Basinand Range province, Trans-Pecos Texas *in* Faulds, J.E. And J.H. Stewart, eds., Accommodation Zones and Transfer Zones: The Regional Segmentation of the Basin and Range Province, Boulder, Colorado, Geological Society of America Special Paper 323.

125 The main source if information on the Presidio Basin is: Mraz, J.R. and Keller, G.R., 1980, Structure of the Presidio bolson area, Texas, interpreted from gravity data, Geological Circular 80-13, Bureau of Economic Geology, Austin.

133 Height of basin fill in Madera Canyon: Henry Geology of Big Bend Ranch State Park, p.66.
Fingers Formation gravels are found at 4,200 feet near the Sotol Vistas viewpoint on Burro Mesa.

144 Calderas and Mineralization: Volcanic Geology and Mineralization in the Chinati Caldera Complex, Trans-Pecos Texas, by T. W. Duex and C. D. Henry. 14 p., 6 figs., 1 table, 1981.

145 The age of the Red Hill copper prospect comes from Gilmer, A.K., Kyle, J.R., Connely, J.N., Mathur, R.D. and C.D. Henry, 1991, The Red Hills intrusive system, Presidio County; the easternmost Laramide porphyry copper-molybdenum deposit in Southwestern North America, *Geology*; May 2003; v. 31; no. 5; p. 447-450.

151 Cepeda, J.C., 1979, The Chinati Mountains Caldera, Presidio County, Texas *in* Cenozoic Geology of the Trans-Pecos Volcanic Field of Texas, Guidebook 19, Bureau of Economic Geology, eds., Walton, A.W. and C.D. Henry.

162 *Chronicles of the Big Bend, A Photographic Memoir of Life on the Border,* by W. D. Smithers, Texas A&M Press.

165 Information on the geology of the Paisano Volcano is from: Parker, D.F., The Paisano Volcano: Stratigraphy, age, and petrogenesis *in* Walton, A.W. and C.D. Henry, eds., Cenozoic geology of the Trans-Pecos volcanic field of Texas: Bureau of Economic Geology, Austin, Guidebook 19, pp. 97-105.

# Glossary

**alluvial fan** A low gently-sloping mass of loose *alluvium,* shaped like a fan, left by a stream at the point where it comes out of a narrow mountain valley into a plain or broad valley.

**alluvium** Sand, clay, silt or gravel deposited recently and unconsolidated, i.e. not cemented together.

**anticline** A fold, generally convex up, whose core contains older rocks.

**ash-flow tuff** *Tuff* created from an ash flow, a mixture of volcanic gases and particles, usually hot, that flows out from explosive viscous *magma* in a volcanic fissure or crater; synonymous with *pyroclastic flow.*

**basalt** A dark-colored igneous rock composed mainly of calcic plagioclase and pyroxene; the fine-grained equivalent of *gabbro*.

**breccia** A coarse-grained rock made up of angular broken rock fragments held together by mineral cement or in a fine-grained matrix. A breccia differs from a *conglomerate* in that the fragments have sharp edges and unworn corners.

**caldera** A large basin-shaped volcanic depression.

**calcite** The principal component of limestone, calcium carbonate $CaCO_3$.

**claystone** A weakly *indurated, sedimentary* rock made up of mainly clay particles i.e. those with diameters of less than 0.01 millimeter.

**collapse breccia** *Breccia* formed by the collapse of rock overlying an opening such as a *magma* chamber or *intrusion*.

**collapse caldera** A *caldera* created by the collapse of a *magma* chamber through the removal of *magma* by volcanic explosions or lava eruptions, or by the removal of *magma* through subterranean pathways.

**conglomerate** A coarse-grained *sedimentary* rock, composed of rounded or sub-angular fragments of rock larger than 2 mm in diameter, set in a fine-grained matrix of sand or silt and commonly cemented by calcium carbonate, iron oxides, silica or hardened clay.

**crystalline** A rock composed of minerals that have a crystalline structure and contain no glass.

**cuesta** A hill or ridge with a gentle slope conforming to the bed or beds that form it on one side and a steep slope on the other formed by outcrops of resistant rocks, the formation of the ridge being controlled by the *differential erosion* of the gently inclined strata.

**dike** A broad, tabular *igneous intrusion* that cuts across the bedding or foliation of the rock into which it was intruded.

**differential erosion** Erosion that occurs at varying rates caused by differences in the hardness or resistance of rocks; softer or weaker rocks are eroded more quickly than harder or more resistant rocks.

**extrusive rock** An *igneous* rock formed from *magma* that has erupted onto the surface of the earth; includes *lavas*, *pyroclastic flows* and volcanic ash.

**fault scarp** A steep slope or cliff formed by movement along a fault and corresponding to the exposed surface of the fault before it was modified by weathering or erosion.

**gabbro** A dark-colored *plutonic igneous* rock composed mainly of calcic feldspar and augite, the approximate equivalent of *basalt*.

**graben** A through or basin bounded on both sides by normal faults dipping into the graben and which has moved down relative to the adjoining fault blocks. See also *horst.*

**granite** A quartz-bearing *felsic plutonic* rock, the intrusive equivalent of rhyolite.

**groundmass** The material between *phenocrysts* in a *porphyritic igneous* rock. It is finer grained than the *phenocrysts* and may be crystalline, glassy or both.

**half graben** A basin bounded on one side by a normal fault.

**hoodoo** A column, pillar or pinnacle of rock created in an region of sporadic heavy rainfall by differential erosion of horizontal strata of differing resistance to weathering. Often occurs in grotesque or eccentric forms.

**horst** A block of the earth's crust that is bounded on opposite sides by faults dipping away from the block and has moved upward relative to the two adjoining blocks. See also *graben.*

**igneous rock** A rock made from molten or partly molten material, i.e. *magma*, that has cooled and solidified.

**indurated** Said of a rock hardened or consolidated by pressure, cementation or heat.

**intrusion** A rock that has formed from a *magma* that has intruded into pre-existing rock.

**laccolith** An *igneous intrusion* parallel to the bedding into which it was intruded except for its roof which is domed.

**lahar** A mudflow, consisting mainly of volcanic rocks and debris, on the flanks of a volcano.

**lava** *Magma* that comes to the earth's surface through a volcanic vent or fissure.

**microgranular** A carbonate sedimentary rock that is minutely granular with particles between 10-60 microns and in which a finer clay-sized matrix is absent.

**mafic** Said of an *igneous* rock that is composed of minerals rich in iron and magnesium, a dark-colored rock. It is the complement of *felsic*.

**megafossil** A fossil large enough to be studied without the aid of a microscope.

**magma** Naturally occurring mobile rock material, generated within the earth and capable of being extruded and intruded, from which *igneous* rocks are derived through cooling. It may or may not contain suspended solids such as crystals and rock fragments.

**microcrystalline** Said of a rock in which the crystals are too small to be seen by the naked eye.

**monocline** A flexure or local steepening of strata.

**nepheline** An *alkali* silicate of the feldspathoid group $(Na,K)[AlSiO_4]$.

**nepheline syenite** A plutonic rock composed of *alkali feldspar* and *nepheline* and perhaps *alkali mafic* minerals; the intrusive equivalent of phonolite (see diagram on page 18).

**normal fault** A fault in which the hanging or upper wall has dropped relative to the foot or lower wall.

**outlier** An area or group of rocks surrounded by older rocks.

**pediment** A gently-sloping broad erosion surface in a semi-arid region at the base of an abrupt and receding mountain front or plateau escarpment.

**phenocryst** A relatively large crystal in a *porphyritic* rock.

**plug** A vertical *igneous* pipe, the channel by which *magma* reached a *volcanic vent.*

**pluton** An *igneous intrusion* formed at depth, of area greater than 40 square miles and with no known floor.

**porphyritic** Describes a rock in which larger crystals are set in a finer grained *groundmass*.

**pyroclastic flow** A flow of *pyroclastic* particles, usually very hot; synonymous with *ash flow.*

**pyroclast** An individual particle of rock ejected by an explosive volcanic eruption; from the Ancient Greek words for *fire* and *broken into pieces*; the adjective is **pyroclastic.**

**rhyolite** A group of fine-grained *extrusive* rocks, typically *porphyritic* and commonly exhibiting flow structures, with *phenocrysts* of quartz and *alkali feldspar* in a glassy to *cryptocrystalline groundmass*, the fine-grained equivalent of *granite*. Rhyolite grades into *trachyte* with a decrease in quartz content.

**sediment** Fragments that originate by weathering of older rocks and forms in layers on the Earth's surface as sand, gravel, silt, mud etc.

**sedimentary rock** A rock that has formed from the consolidation of loose sediment such as fragments of older rock, chemically precipitated material, volcanic pyroclastic fragments, and organic remains.

**sill** A broad, tabular *igneous intrusion* that parallels the bedding or foliation of the rock into which it was intruded.

**syenite** A group of *plutonic* rocks containing *alkali feldspar*, plagioclase, one or more *mafic* minerals, with quartz, if present, only as an accessory; the intrusive equivalent of *trachyte*; with increasing quartz, grades into *granite*.

**syncline** A fold, generally concave up, whose core contains younger rocks..

**thrust fault** A fault where the upper or hanging wall has moved over the lower or foot wall, shortening the earth's crust.

**trachyte** A group of fine-grained *extrusive* rocks, generally *porphyritic*, having *alkali feldspar* and minor *mafic* minerals as the main components, the *extrusive* equivalent of *syenite. Trachyte* grades into *rhyolite* as the quartz content increases.

**tuff** A rock composed of consolidated or cemented volcanic ash; includes *ash-flow tuff* and *ash-fall tuff.*

**volcanic breccia** A *volcaniclastic* rock composed mainly of volcanic fragments greater than 2 mm in diameter.

**volcanic neck** The volcanic rock filling the vent of an extinct volcano.

**volcaniclastic** Pertaining to a *sedimentary* rock composed of mainly or partly of broken fragments of volcanic origin.

**welded tuff** A rock composed of *pyroclasts* welded together by a combination of heat of the particles, the weight of overlying material and hot gases.

# Index